Introduction to Combinatorial Testing

Chapman & Hall/CRC Innovations in Software Engineering and Software Development

Series Editor
Richard LeBlanc
Chair, Department of Computer Science and Software Engineering, Seattle University

AIMS AND SCOPE

This series covers all aspects of software engineering and software development. Books in the series will be innovative reference books, research monographs, and textbooks at the undergraduate and graduate level. Coverage will include traditional subject matter, cutting-edge research, and current industry practice, such as agile software development methods and service-oriented architectures. We also welcome proposals for books that capture the latest results on the domains and conditions in which practices are most effective.

PUBLISHED TITLES

Software Development: An Open Source Approach
Allen Tucker, Ralph Morelli, and Chamindra de Silva

Building Enterprise Systems with ODP: An Introduction to Open Distributed Processing
Peter F. Linington, Zoran Milosevic, Akira Tanaka, and Antonio Vallecillo

Software Engineering: The Current Practice
Václav Rajlich

Fundamentals of Dependable Computing for Software Engineers
John Knight

Introduction to Combinatorial Testing
D. Richard Kuhn, Raghu N. Kacker, and Yu Lei

CHAPMAN & HALL/CRC INNOVATIONS IN SOFTWARE ENGINEERING AND SOFTWARE DEVELOPMENT

Introduction to Combinatorial Testing

D. Richard Kuhn
Raghu N. Kacker
Yu Lei

CRC Press
Taylor & Francis Group
Boca Raton London New York

CRC Press is an imprint of the
Taylor & Francis Group an **informa** business
A CHAPMAN & HALL BOOK

CRC Press
Taylor & Francis Group
6000 Broken Sound Parkway NW, Suite 300
Boca Raton, FL 33487-2742

Printed on acid-free paper
Version Date: 20130426

International Standard Book Number-13: 978-1-4665-5229-6 (Hardback)

Visit the Taylor & Francis Web site at
http://www.taylorandfrancis.com

and the CRC Press Web site at
http://www.crcpress.com

Contents

Preface

SOFTWARE TESTING HAS ALWAYS FACED a seemingly intractable problem: for real-world programs, the number of possible input combinations can exceed the number of atoms in the ocean, so as a practical matter it is impossible to show through testing that the program works correctly for all inputs. Combinatorial testing offers a (partial) solution. Empirical data show that the number of variables involved in failures is small. Most failures are triggered by only one or two inputs, and the number of variables interacting tails off rapidly, a relationship called the *interaction rule*. Therefore, if we test input combinations for even small numbers of variables, we can provide very strong testing at low cost. As always, there is no "silver bullet" answer to the problem of software assurance, but combinatorial testing has grown rapidly because it works in the real world.

This book introduces the reader to the practical application of combinatorial methods in software testing. Our goal is to provide sufficient depth that readers will be able to apply these methods in their own testing projects, with pointers to freely available tools. Included are detailed explanations of how and why to use various techniques, with examples that help clarify concepts in all chapters. Sets of exercises or questions and answers are also included with most chapters. The text is designed to be accessible to an undergraduate student of computer science or engineering, and includes an extensive set of references to papers that provide more depth on each topic. Many chapters introduce some of the theory and mathematics of combinatorial methods. While this material is needed for thorough knowledge of the subject, testers can apply the methods using tools (many freely available and linked in the chapters) that encapsulate the theory, even without in-depth knowledge of the underlying mathematics.

We have endeavored to be as prescriptive as possible, but experienced testers know that standardized procedures only go so far. Engineering judgment is as essential in testing as in development. Because analysis of the input space is usually the most critical step in testing, we have devoted roughly a third of the book to it, in Chapters 3 through 6. It is in this phase that experience and judgment have the most bearing on the success of a testing project. Analyzing and modeling the input space is also a task that is easy to do poorly, because it is so much more complex than it first appears. Chapters 5 and 6 introduce systematic methods for dealing with this problem, with examples to illustrate the subtleties that make the task so challenging to do right.

Chapters 7 through 9 are central to another important theme of this book—combinatorial methods can be applied in many ways during the testing process, and can improve conventional test procedures not designed with these methods in mind. That is, we do not have to completely re-design our testing practices to benefit from combinatorial methods. Any test suite, regardless of how it is derived, provides some level of combinatorial coverage, so one way to use the methods introduced in this book is to create test suites using an organization's conventional procedures, measure their combinatorial coverage, and then supplement them with additional tests to detect complex interaction faults.

The oracle problem—determining the correct output for a given test—is covered in Chapters 10 and 11. In addition to showing how formal models can be used as test oracles, Chapter 11 introduces an approach to integrating testing with formal specifications and proofs of properties by model checkers. Chapters 12 through 15 introduce advanced topics that can be useful in a wide array of problems. Except for the first four chapters, which introduce core terms and techniques, the chapters are designed to be reasonably independent of each other, and pointers to other sections for additional information are provided throughout.

The project that led to this book developed from joint research with Dolores Wallace, and we are grateful for that work and happy to recognize her contributions to the field of software engineering. Special thanks are due to Tim Grance for early and constant support of the combinatorial testing project. Thanks also go to Jim Higdon, Jon Hagar, Eduardo Miranda, and Tom Wissink for early support and evangelism of this work. Donna Dodson, Ron Boisvert, Geoffrey McFadden, David Ferraiolo, and Lee Badger at NIST (U.S. National Institute of Standards and Technology) have been strong advocates for this work. Jon Hagar provided many recommendations for improving the text. Mehra Borazjany, Michael Forbes, Itzel Dominguez Mendoza, Tony Opara, Linbin Yu, Wenhua Wang, and

Laleh SH. Ghandehari made major contributions to the software tools developed in this project. We have benefitted tremendously from interactions with researchers and practitioners, including Bob Binder, Paul Black, Renee Bryce, Myra Cohen, Charles Colbourn, Howard Deiner, Elfriede Dustin, Mike Ellims, Al Gallo, Vincent Hu, Justin Hunter, Greg Hutto, Aditya Mathur, Josh Maximoff, Carmelo Montanez-Rivera, Jeff Offutt, Vadim Okun, Michael Reilly, Jenise Reyes Rodriguez, Rick Rivello, Sreedevi Sampath, Itai Segall, Mike Trela, Sergiy Vilkomir, and Tao Xie. We also gratefully acknowledge NIST SURF (Summer Undergraduate Research Fellowships) students Kevin Dela Rosa, William Goh, Evan Hartig, Menal Modha, Kimberley O'Brien-Applegate, Michael Reilly, Malcolm Taylor, and Bryan Wilkinson who contributed to the software and methods described in this document. We are especially grateful to Randi Cohen, editor at Taylor & Francis, for making this book possible and for timely guidance throughout the process. Certain software products are identified in this document, but such identification does not imply recommendation by the U.S. National Institute for Standards and Technology, nor does it imply that the products identified are necessarily the best available for the purpose.

Authors

D. Richard Kuhn is a computer scientist in the Computer Security Division of the National Institute of Standards and Technology (NIST). He has authored or coauthored more than 100 publications on information security, empirical studies of software failure, and software assurance, and is a senior member of the IEEE. He co-developed the role-based access control model (RBAC) used throughout the industry and led the effort establishing RBAC as an ANSI (American National Standards Institute) standard. Before joining NIST, he worked as a systems analyst with NCR Corporation and the Johns Hopkins University Applied Physics Laboratory. He earned an MS in computer science from the University of Maryland College Park, and an MBA from the College of William & Mary.

Raghu N. Kacker is a senior researcher in the Applied and Computational Mathematics Division (ACMD) of the Information Technology Laboratory (ITL) of the U.S. National Institute of Standards and Technology (NIST). His current interests include software testing and evaluation of the uncertainty in outputs of computational models and physical measurements. He has a PhD in statistics and has coauthored over 100 refereed papers. Dr. Kacker has been elected Fellow of the American Statistical Association and a Fellow of the American Society for Quality.

Yu Lei is an associate professor in Department of Computer Science and Engineering at the University of Texas, Arlington. He earned his PhD from North Carolina State University. He was a member of the Technical Staff in Fujitsu Network Communications, Inc. from 1998 to 2002. His current research interests include automated software analysis and testing, with a special focus on combinatorial testing, concurrency testing, and security testing.

Note of Appreciation

Two people who have made major contributions to the methods introduced in this book are James Lawrence of George Mason University, and James Higdon of Jacobs Technology, Eglin AFB. Jim Lawrence has been an integral part of the team since the beginning, providing mathematical advice and support for the project. Jim Higdon was co-developer of the sequence covering array concept described in Chapter 10, and has been a leader in practical applications of combinatorial testing.

If no additional faults are detected by 4-way tests, then it may be reasonable to conduct additional testing only for 4-way interactions, but if more ... the greatest ... [illegible] ... set of software failures, the number of faults detected has declined monotonically with t, and the DOM testing results are consistent with the earlier finding. Following this strategy for the DOM testing would result in running five-way tests, though ... then stopping because no additional faults were detected beyond four-way testing. Alternatively, given the [illegible] ... [illegible] ... [illegible] ... [illegible] ... [illegible] ... four-way testing ... [illegible] ... The results of ... [illegible] ... [illegible] ...

2.? RICH WEB APPLICATIONS

Chad M. Maughan

Rich web applications (RWA) are rapidly gaining in popularity. RWAs focus on providing a rich user experience within browsers and often do not rely on proprietary client-side browser plug-ins such as Adobe Flash. Typically, an RWA consists of a single HTML page that acts as an entry point. AJAX calls to a stateless (i.e., sessionless) server retrieve data that is then used to dynamically manipulate the DOM. Semantic (i.e., human understandable) URL fragments control history and browser interactions. Client-side JavaScript frameworks such as Google Web Toolkit (GWT), Backbone.js, Angular.js, and Ember.js facilitate the movement of logic and processing from the server side to the client side in the browser while helping to deliver a rich user experience.

While end users may benefit from the migration of computation logic from the server to the browser, new testing challenges arise. The exceptions and coding irregularities no longer occur solely on the server in a predictable, controlled, and centralized location. Errors now occur in a client's browser ... [illegible] types and versions of browsers, computers, and environments in a dynamically typed language that is difficult to statically analyze. Maughan [129] proposes the use of combinatorial testing for both server and client side

Nomenclature

n = number of variables or parameters in tests

t = interaction strength; number of variables in a combination

N = number of tests

v_i = number of values for variable i

CA(N, n, v, t) = t-way covering array of N rows for n variables with v values each

CAN(t, n, v) = number of rows in a t-way covering array of n variables with v values each

OA(N, v^k, t) = t-way orthogonal array of entries from the set {0, 1, ..., (v – 1)}

$C(n,t) = \binom{n}{t} = \frac{n!}{t!(n-t)!}$, the number of t-way combinations of n parameters

SCA(N, S, t) = an $N \times S$ sequence covering array where entries are from a finite set S of symbols, such that every t-way permutation of symbols from S occurs in at least one row

(p, t)-completeness = proportion of the $C(n, t)$ combinations in a test array of n rows that have configuration coverage of at least p

Φ_t = proportion of combinations with full t-way variable-value configuration coverage

M_t = minimum t-way variable value configuration coverage for a test set

S_t = proportion of t-way variable value configuration coverage for a test set

F = set of t-way combinations in failing tests

P = set of t-way combinations in passing tests

F^+ = augmented set of combinations in failing tests

P^+ = augmented set of combinations in passing tests

CHAPTER 1

Combinatorial Methods in Testing

DEVELOPERS OF LARGE SOFTWARE systems often notice an interesting phenomenon: usage of an application suddenly increases, and components that worked correctly for years develop previously undetected failures. For example, the application may have been installed with a different operating system (OS) or database management system (DBMS) than used previously, or newly added customers may have account records with combinations of values that have not occurred before. Some of these rare combinations trigger failures that have escaped previous testing and extensive use. Such failures are known as *interaction failures*, because they are only exposed when two or more input values interact to cause the program to reach an incorrect result.

1.1 SOFTWARE FAILURES AND THE INTERACTION RULE

Interaction failures are one of the primary reasons why software testing is so difficult. If failures only depended on one variable value at a time, we could simply test each value once, or for continuous-valued variables, one value from each representative range or equivalence class. If our application had inputs with v values each, this would only require a total of v tests—one value from each input per test. Unfortunately, the real world is much more complicated than this.

Combinatorial testing can help detect problems like those described above early in the testing life cycle. The key insight underlying t-way combinatorial testing is that not every parameter contributes to every failure and most failures are triggered by a single parameter value or interactions between a relatively small number of parameters. For example, a router may be observed to fail only for a particular protocol when packet volume exceeds a certain rate, a 2-way interaction between protocol type and packet rate. Figure 1.1 illustrates how such a 2-way interaction may happen in code. Note that the failure will only be triggered when both *pressure* < *10* and *volume* > *300* are true. To detect such interaction failures, software developers often use "pairwise testing," in which all possible pairs of parameter values are covered by at least one test. Its effectiveness results from the fact that most software failures involve only one or two parameters.

Pairwise testing can be highly effective and good tools are available to generate arrays with all pairs of parameter value combinations. But until recently only a handful of tools could generate combinations beyond 2-way, and most that did could require impractically long times to generate 3-way, 4-way, or 5-way arrays because the generation process is mathematically complex. Pairwise testing, that is, 2-way combinations, is a common approach to combinatorial testing because it is computationally tractable and reasonably effective.

But what if some failure is triggered only by a very unusual combination of 3, 4, or more values? It is unlikely that pairwise tests would detect this unusual case; we would need to test 3- and 4-way combinations of values. But is testing all 4-way combinations enough to detect all errors? It is important to understand the way in which interaction failures occur in real systems, and the number of variables involved in these failure triggering interactions.

```
if (pressure < 10) {
     // do something
     if (volume > 300)   {
          faulty code!  BOOM!
     }
     else {
          good code, no problem
     }
}
else {
     // do something else
}
```

FIGURE 1.1 2-Way interaction failures are triggered when two conditions are true.

What degree of interaction occurs in real failures in real systems? Surprisingly, this question had not been studied when the National Institute of Standards and Technology (NIST) began investigating interaction failures in 1999. An analysis of 15 years of medical device recall data [212] included an evaluation of fault-triggering combinations and the testing that could have detected the faults. For example, one problem report said that "if device is used with old electrodes, an error message will display, instead of an equipment alert." In this case, testing the device with old electrodes would have detected the problem. Another indicated that "upper limit CO_2 alarm can be manually set above upper limit without alarm sounding." Again, a single test input that exceeded the upper limit would have detected the fault. Other problems were more complex. One noted that "if a bolus delivery is made while pumps are operating in the body weight mode, the middle LCD fails to display a continual update." In this case, detection would have required a test with the particular pair of conditions that caused the failure: bolus delivery while in body weight mode. One description of a failure manifested on a particular pair of conditions was "the ventilator could fail when the altitude adjustment feature was set on 0 meters and the total flow volume was set at a delivery rate of less than 2.2 liters per minute." The most complex failure involved four conditions and was presented as "the error can occur when demand dose has been given, 31 days have elapsed, pump time hasn't been changed, and battery is charged."

Reviews of failure reports across a variety of domains indicated that all failures could be triggered by a maximum of 4-way to 6-way interactions [103–105,212] for the applications studied. As shown in Figure 1.2, the detection rate increased rapidly with interaction strength (the interaction level t in t-way combinations is often referred to as *strength*). With the NASA application, for example, 67% of the failures were triggered by only a single parameter value, 93% by 2-way combinations, and 98% by 3-way combinations. The detection rate curves for the other applications studied are similar, reaching 100% detection with 4-way to 6-way interactions. Studies by other researchers [14,15,74,222] have been consistent with these results.

Failures appear to be caused by interactions of only a few variables, so tests that cover all such few-variable interactions can be very effective.

These results are interesting because they suggest that, while pairwise testing is not sufficient, the degree of interaction involved in failures is

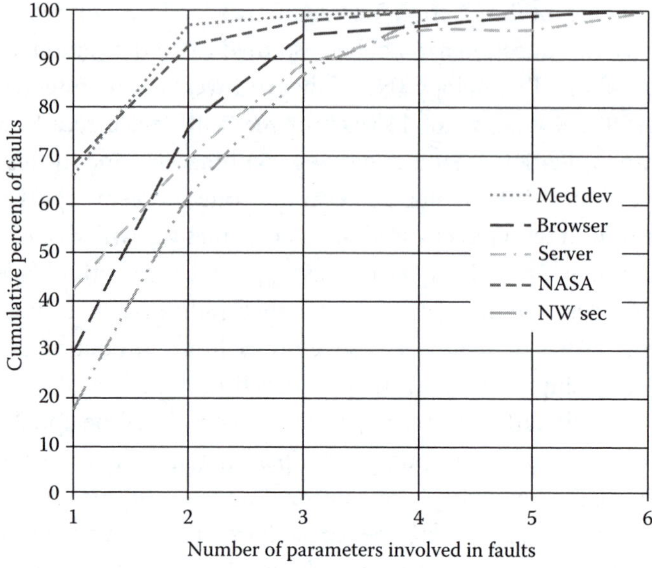

FIGURE 1.2 **(See color insert.)** The Interaction Rule: Most failures are triggered by one or two parameters interacting, with progressively fewer by 3, 4, or more.

relatively low. We summarize this result in what we call the *interaction rule*, an empirically derived [103–105] rule that characterizes the distribution of interaction faults:

> **Interaction Rule**: Most failures are induced by single factor faults or by the joint combinatorial effect (interaction) of two factors, with progressively fewer failures induced by interactions between three or more factors.

The maximum degree of interaction in actual real-world faults so far observed is six. This is not to say that there are no failures involving more than six variables, only that the available evidence suggests they are rare (more on this point below). Why is the interaction rule important? Suppose we somehow know that for a particular application, any failures can be triggered by 1-way, 2-way, or 3-way interactions. That is, there are some failures that occur when certain sets of two or three parameters have particular values, but no failure that is only triggered by a 4-way interaction. In this case, we would want a test suite that covers all 3-way combinations of parameter values (which automatically guarantees 2-way coverage as well). If there are some 4-way interactions that are not covered, it will not matter from a fault detection standpoint, because none of the failures involve 4-way interactions. Therefore in this example, covering all 3-way combinations is in a certain sense equivalent to exhaustive testing. It will not test all possible inputs, but those inputs that are not tested would not make any difference in finding faults in the software. For this reason, we sometimes refer to this approach as "pseudo-exhaustive" [103], analogous to the digital circuit testing method of the same name [131,200]. The obvious flaw in this scenario is our assumption that we "somehow know" the maximum number of parameters involved in failures. In the real world, there may be 4-way, 5-way, or even more parameters involved in failures, so our test suite covering 3-way combinations might not detect them. But if we can identify a practical limit for the number of parameters in combinations that must be tested, and this limit is not too large, we may actually be able to achieve the "pseudo-exhaustive" property. This is why it is essential to understand interaction faults that occur in typical applications.

Some examples of such interactions were described previously for medical device software. To get a better sense of interaction problems in real-world software, let us consider some examples from an analysis of over 3000 vulnerabilities from the National Vulnerability Database, which is a collection of all publicly reported security issues maintained by NIST and the Department of Homeland Security:

- Single variable (1-way interaction): Heap-based *buffer_overflow* in the SFTP protocol handler for Panic Transmit … allows remote attackers to execute arbitrary code via a long ftps:// URL.
- 2-Way interaction: *Single character search string* in conjunction with a *single character replacement string*, which causes an "off by one overflow."
- 3-Way interaction: Directory traversal vulnerability when *register_globals is enabled* and *magic_quotes is disabled* and.. *(dot dot) in the page parameter.*

The single-variable case is a common problem: someone forgot to check the length of an input string, allowing an overflow in the input buffer. A test set that included any test with a sufficiently long input string would have detected this fault. The second case is more complex, and would not necessarily have been caught by many test suites. For example, a requirements-based test suite may have included tests to ensure that the software was capable of accepting search strings of 1 to N characters, and others to check the requirement that 1 to N character replacement strings could be entered. But unless there was a single test that included *both* a one-character search string and a one-character replacement string, the application could have passed the test suite without detecting the problem. The 3-way interaction example is even more complex, and it is easy to see that an *ad hoc* requirements-based test suite might be constructed without including a test for which all three of the italicized conditions were true. One of the key features of combinatorial testing is that it is designed specifically to find this type of complex problem, despite requiring a relatively small number of tests.

As discussed above, an extensive body of empirical research suggests that testing 2-way (pairwise) combinations is not sufficient, and a significant proportion of failures result from 3-way and higher strength interactions. This is an important point, since the only combinatorial method many testers are familiar with is pairwise/2-way testing, mostly because good algorithms to produce 3-way and higher strength tests were not available. Fortunately, better algorithms and tools now make high strength t-way tests possible, and one of the key research questions in this field is thus: What t-way combination strength interaction is needed to detect all interaction failures? (Keep in mind that not all failures *are* interaction failures—many result from timing considerations, concurrency problems, and other factors that are not addressed by conventional combinatorial testing.) As we have discussed, failures seen thus far in real-world systems seem to involve six or fewer parameters interacting. However, it is *not* safe to assume that there are no software failures involving 7-way or higher interactions. It is likely that there are some that simply have not been recognized. One can easily construct an example that could escape detection by t-way testing for any arbitrary value of t, by creating a complex conditional with $t + 1$ variables:

```
if (v1 && ... && vt && vt+1) {/* bad code */}.
```

In addition, analysis of the branching conditions in avionics software shows up to 19 variables in some cases [42]. Experiments on using combinatorial testing to achieve code coverage goals such as line, block, edge, and condition coverage have found that the best coverage was obtained with 7-way combinations [163,188], but code coverage is not the same as fault detection. Our colleague Linbin Yu has found up to 9-way interactions in some conditional statements in the Traffic Collision Avoidance System software [216] that is often used in testing research, although 5-way combinations were sufficient to detect all faults in this set of programs [103] (*t*-way tests always include some higher strength combinations, or the 9-way faults may also have been triggered by <9 variables). Because the number of branching conditions involving *t* variables decreases rapidly as *t* increases, it is perhaps not surprising that the number of failures decreases as well. The available empirical research on this issue is covered in more detail in a web page that we maintain [143], and summarized in Appendix B. Because failures involving more than six parameters have not been observed in fielded software, most combinatorial testing tools generate up to 6-way arrays.

Because of the interaction rule, ensuring coverage of all 3-way, possibly up to 6-way combinations may provide high assurance. As with most issues in software, however, the situation is not that simple. Efficient generation of test suites to cover all *t*-way combinations is a difficult mathematical problem that has been studied for nearly a century, although recent advances in algorithms have made this practical for most testing. An additional complication is that most parameters are continuous variables which have possible values in a very large range ($\pm 2^{31}$ or more). These values must be discretized to a few distinct values. Most glaring of all is the problem of determining the correct result that should be expected from the system under test (SUT) for each set of test inputs. Generating 1000 test data inputs is of little help if we cannot determine what SUT should produce as output for each of the 1000 tests.

With the exception of covering combinations, these challenges are common to all types of software testing, and a variety of good techniques have been developed for dealing with them. What has made combinatorial testing practical today is the development of efficient algorithms to generate tests covering *t*-way combinations, and effective methods of integrating the tests produced into the testing process. A variety of approaches introduced in this book can be used to make combinatorial testing a practical and effective addition to the software tester's toolbox.

Advances in algorithms have made combinatorial testing beyond pairwise finally practical.

Notes on terminology: we use the definitions below, following the Institute of Electrical and Electronics Engineers (IEEE) Glossary of Terms [97]. The term "bug" may also be used where its meaning is clear.

Error: A mistake made by a developer. This could be a coding error or a misunderstanding of requirements or specification.

Fault: A difference between an incorrect program and one that correctly implements a specification. An error may result in one or more faults.

Failure: A result that differs from the correct result as specified. A fault in code may result in zero or more failures, depending on inputs and execution path.

The acronym SUT (system under test) refers to the target of testing. It can be a function, a method, a complete class, an application, or a full system including hardware and software. Sometimes, a SUT is also referred as a test object (TO) or artifact under test (AUT). That is, SUT is not meant to imply only the system testing phase.

1.2 TWO FORMS OF COMBINATORIAL TESTING

There are basically two approaches to combinatorial testing—use combinations of *configuration* parameter values, or combinations of *input parameter* values. In the first case, we select combinations of values of configurable parameters. For example, a server might be tested by setting up all 4-way combinations of configuration parameters such as number of simultaneous connections allowed, memory, OS, database size, DBMS type, and others, with the same test suite run against each configuration. The tests may have been constructed using any methodology, not necessarily combinatorial coverage. The combinatorial aspect of this approach is in achieving combinatorial coverage of all possible t-way configuration parameter values. (Note that the terms *variable* and *factor* are often used interchangeably with *parameter* to refer to inputs to a function or a software program.)

Combinatorial testing can be applied to configurations, input data, or both.

In the second approach, we select combinations of *input data* values, which then become part of complete test cases, creating a test suite for the application. In this case, combinatorial coverage of input data values is required for tests constructed. A typical *ad hoc* approach to testing involves subject matter experts setting up use case scenarios, then selecting input values to exercise the application in each scenario, possibly supplementing these tests with unusual or suspected problem cases. In the combinatorial approach to input data selection, a test data generation tool is used to cover all combinations of input values up to some specified limit. One such tool is automated combinatorial testing for software (ACTS) (described in Appendix C), which is available freely from NIST.

Aspects of both configuration testing and input parameter testing may appear in a great deal of practical testing. Both types may be applied for thorough testing, with combinations of input parameters applied to each configuration combination. In state machine approaches (Chapter 6), other variations appear—parameters are inputs that may determine the presence or absence of other parameters, or both program variables and states may be treated as test parameters. But a wide range of testing problems can be categorized as either configuration or input testing, and these approaches are analyzed in more detail in later chapters.

1.2.1 Configuration Testing

Many, if not most, software systems have a large number of configuration parameters. Many of the earliest applications of combinatorial testing were in testing all pairs of system configurations. For example, telecommunications software may be configured to work with different types of call (local, long distance, international), billing (caller, phone card, 800), access (ISDN, VOIP, PBX), and server for billing (Windows Server, Linux/MySQL, Oracle). The software must work correctly with all combinations of these, so a single test suite could be applied to all pairwise combinations of these four major configuration items. Any system with a variety of configuration options is a suitable candidate for this type of testing.

For example, suppose we had an application that is intended to run on a variety of platforms comprised of five components: an operating system (Windows XP, Apple OS X, Red Hat Enterprise Linux), a browser (Internet Explorer, Firefox), protocol stack (IPv4, IPv6), a processor (Intel, AMD), and a database (MySQL, Sybase, Oracle), a total of $3 \times 2 \times 2 \times 2 \times 3 = 72$ possible platforms. With only 10 tests, shown in Table 1.1, it is possible to test every component interacting with every other component at least once, that is, all possible pairs of platform components are covered. While this gain in efficiency—10 tests instead of 72—is respectable, the improvement for larger test problems can be spectacular, with 2- and 3-way tests often requiring <1% of the tests needed for exhaustive testing. In general, the larger the problem, the greater the efficiency gain from combinatorial testing.

TABLE 1.1 Pairwise Test Configurations

Test	OS	Browser	Protocol	CPU	DBMS
1	XP	IE	IPv4	Intel	MySQL
2	XP	Firefox	IPv6	AMD	Sybase
3	XP	IE	IPv6	Intel	Oracle
4	OS X	Firefox	IPv4	AMD	MySQL
5	OS X	IE	IPv4	Intel	Sybase
6	OS X	Firefox	IPv4	Intel	Oracle
7	RHEL	IE	IPv6	AMD	MySQL
8	RHEL	Firefox	IPv4	Intel	Sybase
9	RHEL	Firefox	IPv4	AMD	Oracle
10	OS X	Firefox	IPv6	AMD	Oracle

1.2.2 Input Testing

Even if an application has no configuration options, some form of input will be processed. For example, a word-processing application may allow the user to select 10 ways to modify some highlighted text: *subscript, superscript, underline, bold, italic, strikethrough, emboss, shadow, small caps*, or *all caps*. The font-processing function within the application that receives these settings as input must process the input and modify the text on the screen correctly. Most options can be combined, such as bold and small caps, but some are incompatible, such as subscript and superscript.

Thorough testing requires that the font-processing function work correctly for all valid combinations of these input settings. But with 10 binary inputs, there are $2^{10} = 1024$ possible combinations. Fortunately, the empirical analysis reported above shows that failures appear to involve a small number of parameters, and that testing all 3-way combinations can often detect 90% or more of bugs. For a word-processing application, testing that detects better than 90% of bugs may be a cost-effective choice, but we need to ensure that all 3-way combinations of values are tested. To do this, or to construct the configuration tests shown in Table 1.1, we create a matrix that covers all t-way combinations of variable values, where $t = 2$ for the configuration problem described previously and $t = 3$ for the 10 binary inputs in this section. This matrix is known as a *covering array* [27,31,51,53,58,97,116,205].

How many t-way combinations must be covered in the array? Consider the example of 10 binary variables. There are C(10, 2) = 45 pairs of variables (*ab, ac, ad,*...). For each pair, the two binary variables can be assigned $2^2 = 4$

possible values: 00, 01, 10, 11. So the number of 2-way combinations that must be covered in the array is $2^2 \times C(10, 2) = 4 \times 45 = 180$. For 3-way combinations, the variables can be assigned eight possible values: 000, 001, 010, ... Selecting three variables can be done in $C(10, 3) = 120$ ways, so there are $2^3 \times C(10, 3) = 960$ possible parameter settings to be covered. In general, there are v^t t-way combinations of v values, so for n parameters we have

$$\mathit{Total\ variable\ settings} = v^t \binom{n}{t}.$$

Generally not all parameters have the same number of test values. In combinatorics parlance, these are referred to as "mixed level" parameters. For n different parameters, where v_{ij} refers to the jth value for the ith parameter, we need to cover:

$$\mathit{Total\ mixed\ level\ combinations} = \textstyle\sum_i v_{i1} \times \cdots \times v_{it}$$

$$\forall i = 1 \ldots \binom{n}{t} \text{ } t\text{-way combinations}$$

As we will see in the next section, a very large number of such combinations can be covered in remarkably few tests. Algorithms to compute covering arrays efficiently have been developed and are now implemented in practical tools.

1.3 COVERING ARRAYS

> The key component is a *covering array,* which includes all *t*-way combinations. Each column is a parameter. Each row is a test.

An example of a covering array is given in Figure 1.3, which shows a 3-way covering array for 10 variables with two values each. The interesting property of this array is that any three columns contain all eight possible values for three binary variables. For example, taking columns *F*, *G*, and *H*, we can see that all eight possible 3-way combinations (000, 001, 010, 011, 100, 101, 110, 111) occur somewhere in the three columns together. In fact, any combination of three columns chosen in any order will also contain all eight possible values. Collectively, therefore, this set of tests will exercise all 3-way combinations of input values in only 13 tests, as compared with 1024 for exhaustive coverage. Similar arrays can

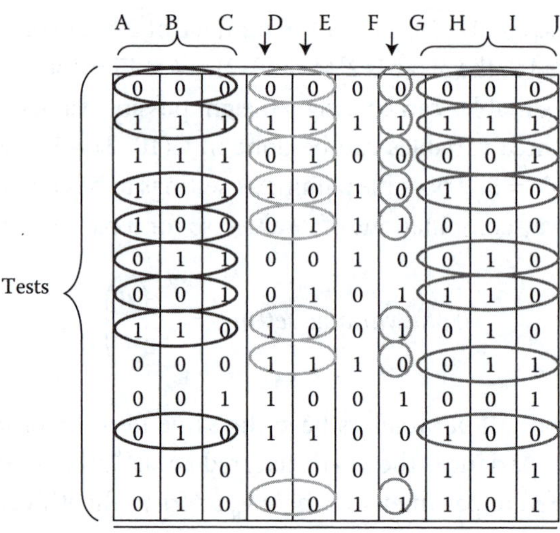

FIGURE 1.3 **(See color insert.)** A 3-way covering array includes all 3-way combinations of values.

be generated to cover all t-way combinations, for whatever value of t is appropriate to the problem.

1.3.1 Covering Array Definition

A covering array CA(N, n, s, t) is an $N \times n$ matrix in which entries are from a finite set S of s symbols such that each $N \times t$ subarray contains each possible t-tuple at least once. For example, in the matrix above, we saw that all eight possible 3-tuples (3-way combinations) of the binary variables occurred at least once. The number t is referred to as the *strength* of the array. A covering array must satisfy the t-covering property: when any t of the k columns are chosen, all v^t of the possible t-tuples must appear among the rows. The "size" of an array is usually given as its number N of rows, where the number of columns is fixed.

This definition can be generalized to the case where k_1 columns have v_1 distinct values, k_2 columns have v_2 distinct values, and so on. A covering array with n_1 columns of v_1 distinct values, n_2 columns of v_2 distinct values, and so on is designated by the exponential expression $v_1^{n1}\ v_2^{n2} \ldots v^{nk}$. Example: An array that has three columns with two distinct values each, two columns with five distinct values each, and four columns with six distinct values each is called a $2^3 5^2 6^4$ array. Note that if the columns represent nine parameters and their input values for a SUT, the number of tests required for exhaustive testing would be $2^3 5^2 6^4 = 259{,}200$ tests. The covering array in Figure 1.3 is a 2^{10} array, since it has 10 columns of binary variables.

1.3.2 Size of Covering Arrays

It is important to understand how covering array size is affected by the attributes of a testing problem to get a sense of how to apply combinatorial testing in practice. Since we are discussing tests and parameters, the notation is a bit different than as used above in the formal definition of a covering array. It has been shown [61] that, in general, the number of rows (tests) for a covering array constructed with a greedy algorithm grows as

$$v^t \log n \tag{1.1}$$

where
v = number of possible values that each variable can take on
t = interaction strength, that is, t-way interactions
n = number of variables or parameters for the tests

When a covering array is produced, the number of tests will be proportional to this expression, not equal to it, but taking a look at the components of this expression will help in understanding how the characteristics of a testing problem affect the number of tests needed. This is a "good news/bad news" situation. The good news is that the number of tests increases only logarithmically with the number of parameters, n. Thus, testing systems with 50 inputs will not require significantly more tests than for 40 inputs. However, the bad news is that the number of tests increases exponentially with t, the interaction strength. So, 4-way testing will be much more expensive than 3-way testing. Note another aspect of the first component, v^t, of expression (1.1). The exponent t applies to v, the number of values that each variable can take on, so the value of v can have an enormous effect on the number of tests.

Since many or most variables will be continuous-valued (within the limitations of digital hardware), values must be discretized from some range of integer or floating point numbers. The input range must be partitioned into a relatively small number of discrete values (see Section 4.1) to keep the number of tests to a minimum. In practice, it is generally a good idea to keep the number of values per variable to 10 or fewer. Figure 1.4 shows the number of tests required for 10 through 100 parameters for various values of v for $t = 2$.

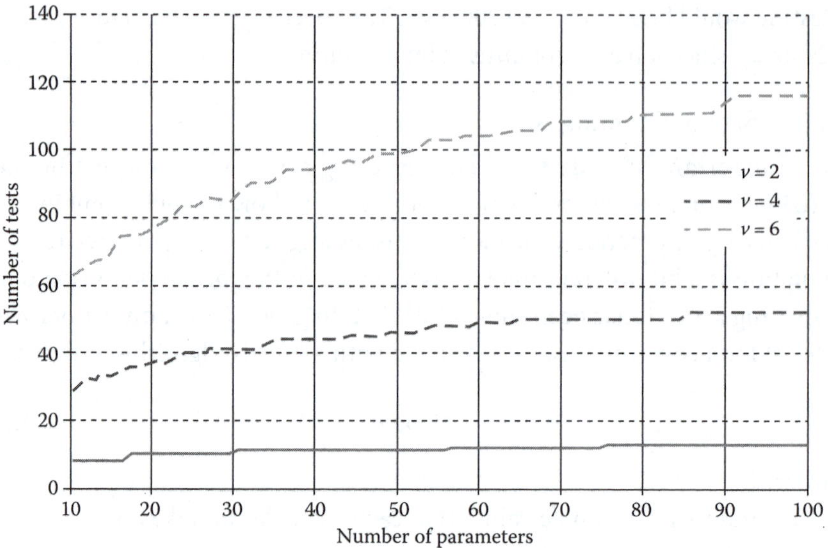

FIGURE 1.4 (**See color insert.**) Number of tests, $t = 2$.

There is no known formula for computing the smallest possible covering array for a particular problem. A database maintained by Charles Colbourn at Arizona State University collects the best-known sizes of covering arrays for a broad range of configurations ranging from $t = 2$ to $t = 6$ (see http://www.public.asu.edu/~ccolbou/src/tabby/catable.html). Many algorithms have been developed for computing covering arrays, but there is no uniformly best algorithm, in the sense of always computing the smallest possible array. Certain algorithms produce very compact arrays for some configurations, but perform poorly on others. More on algorithm design can be found in Chapter 15.

At this point, it is important to point out that covering arrays are not the only way to produce combinatorial coverage. Any test set may cover a large number of parameter value combinations, and ways to measure such coverage are introduced in Chapter 7. As introduced previously in this chapter, the motivation for our interest in combinatorial methods is the empirical observation—the Interaction Rule—that a relatively small number of parameters interact in producing failures in real-world software. We thus want to cover in testing as many combinations as possible, and covering arrays are just one approach (although usually the most efficient). We can measure the combinatorial coverage of just about any test set, regardless of how it is produced. A combinatorial approach to testing is thus compatible with a broad range of test strategies, and this approach can improve testing in a variety of ways that will be introduced in this book.

1.4 THE TEST ORACLE PROBLEM

Even with efficient algorithms to produce covering arrays, the oracle problem remains—testing requires both test data and results that should be expected for each data input. High interaction strength combinatorial testing may require a large number of tests in some cases, although not always. This section summarizes some approaches to solving the oracle problem that are particularly suited to automated or semi-automated combinatorial testing. Note that there are other test oracle methods as well. One of the most widely used approaches is of course to have human experts who analyze test cases and determine the expected results. It is also possible that some or all of the functionality of the SUT will exist in another program. For example, the new code may be modifying one part of an existing program, so old tests may be reused. In some cases, all of the functions may exist in another program whose results can be compared with the SUT, for example, in a version that runs on another platform or a separate implementation of a compiler or network protocol standard. Here, we summarize some approaches for the general case where the SUT presents all or mostly new functionality.

Crash testing: the easiest and least-expensive approach is to simply run tests against the SUT to check whether any unusual combination of input values causes a crash or other easily detectable failure. Execution traces and memory dumps may then be analyzed to determine the cause of the crash. This is similar to the procedure used in some types of "fuzz testing" [189], which sends random values against the SUT. It should be noted that although pure random testing will generally cover a high percentage of t-way combinations, 100% coverage of combinations usually requires a random test set much larger than a covering array. For example, all 3-way combinations of 10 parameters with 4 values each can be covered with 151 tests. A purely random generation requires over 900 tests to provide full 3-way coverage.

Assertions: An increasingly popular technique is to embed assertions within code to ensure proper relationships between data, for example, as preconditions, postconditions, or consistency checks. Tools such as the Java Modeling Language (JML) can be used to introduce very complex assertions, effectively embedding a formal specification within the code. The embedded assertions serve as an executable form of the specification, thus providing an oracle for the testing phase. With embedded assertions, exercising the application with all t-way combinations can provide reasonable assurance that the code works correctly across a very wide range of inputs. This approach has been used successfully for testing smart cards, with embedded JML assertions acting as an oracle for combinatorial tests [66]. Results showed that 80–90% of failures could be found in this way.

Several types of test oracle can be used, depending on resources and the system under test.

Model-based test generation uses a mathematical model of the SUT and a simulator or model checker to generate expected results for each input [1,16,19,133,150]. If a simulator can be used, the expected results can be generated directly from the simulation, but model checkers are widely available and can also be used to prove properties such as liveness in parallel processes, in addition to generating tests. Conceptually, a model checker can be viewed as exploring all states of a system model to determine if a property claimed in a specification statement is true. What makes a model checker particularly valuable is that if the claim is false, the model checker not only reports this, but also provides a "counterexample" showing how the claim can be shown false. If the claim is false, the model checker indicates this and provides a trace of parameter input values and states that will prove it is false. In effect, this is a complete test case, that is, a set of parameter values and expected result. It is then simple to map these values into complete test cases in the syntax needed for the SUT. Chapter 12 develops detailed procedures for applying the model-based test oracle generation.

1.5 QUICK START: HOW TO USE THE BASICS OF COMBINATORIAL METHODS RIGHT AWAY

This book introduces a wide range of topics in combinatorial methods for software testing, sufficient for handling many practical challenges in software assurance. Most testers, however, will not face all of the types of test problems covered in this book, at least not on every project. Many test problems require a core set of methods, possibly with one or two specialized topics. As with many subjects, one of the best ways to approach combinatorial testing is to start small; try the basics to get a feel for how it works, then supplement these methods as needed. This book is designed for such an approach. Readers who are anxious to learn by applying some of the methods introduced here can use the following steps:

1. Read Chapter 1, to learn why combinatorial methods are effective and what to expect.
2. Read Chapters 3 and 4, for step-by-step approaches to input testing and configuration testing (as introduced in Section 1.2).
3. Download and install the Java program ACTS or another covering array tool (see Appendix D).
4. Develop a covering array of tests using ACTS or other tool, then run the tests.

After reading this chapter to understand why combinatorial testing works, readers can also review the two case studies in Chapter 2. These two testing problems are practical examples that illustrate the basics in situations which include many features of web application testing problems. Following the steps above is really just getting started, of course, and readers doing practical testing will encounter problems that can be solved using methods from the other chapters. For example, thorough input domain analysis (Chapters 5 and 6) to select test values is essential for strong testing and assurance. But trying the steps listed above on one of your own small testing problems will likely make the rest of the techniques introduced in the book easier and more interesting to apply.

1.6 CHAPTER SUMMARY

1. Empirical data suggest that software failures are caused by the interaction of relatively few parameter values, and that the proportion of failures attributable to t-way interactions declines very rapidly with increase in t. That is, usually single parameter values or a pair of values will be the cause of a failure, but increasingly smaller proportions are caused by 3-, 4-, and higher-order interactions. This relationship is called the *Interaction Rule.*
2. Because a small number of parameters are involved in failures, we can attain a high degree of assurance by testing all t-way interactions, for an appropriate interaction strength t (2–6 usually). The number of t-way tests that will be required is proportional to $v^t \log n$, for n parameters with v values each.

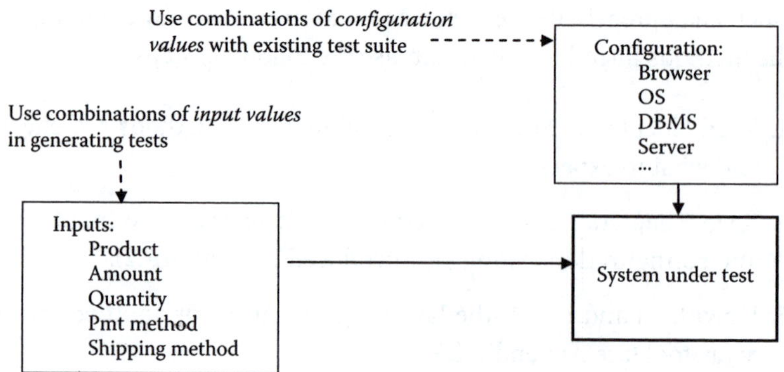

FIGURE 1.5 Combinatorial testing may be used on input values or configurations.

3. A mathematical construct called a *covering array* can be used to produce tests that cover all *t*-way combinations. There is no "best" covering array construction algorithm, in the sense of always producing an optimal array.
4. As with all other types of testing, the oracle problem must be solved—that is, for every test input, the expected output must be determined in order to check if the application is producing the correct result for each set of inputs. A variety of methods can be used to solve the oracle problem.
5. Combinatorial methods can be applied to *configurations* of the SUT or to *input values*, or both. Figure 1.5 contrasts the two approaches to combinatorial testing. With the first approach, we may run the same test set against all *t*-way combinations of configuration options, while for the second approach, we would construct a test suite that covers all *t*-way combinations of input transaction fields. Of course, these approaches could be combined with the combinatorial tests run against all the configuration combinations.

REVIEW

Q: What is the Interaction Rule and why is it significant?

A: The rule states that most failures are caused by one or two parameters interacting, with progressively fewer failures caused by 3, 4, or more parameter interactions. So if all failures are caused by the interaction of a small number of parameters, then testing combinations of these values up to an appropriate level is likely to catch nearly all errors.

Q: What are the two primary ways in which combinatorial testing is applied?

A: Combinatorial testing may be applied to configuration settings or input values, or both.

Q: What is a covering array?

A: A matrix that covers all t-way combinations of variable values, for some specified value of t.

Q: What does the term *strength* refer to in the context of t-way testing?

A: The value of t, the number of test parameters for which all allowed combinations are covered.

Q: Does use of a 4-way covering array guarantee detection of all 4-way faults? Why or why not?

A: No, because in most cases some input space partitioning is required for continuous variables. For example, a "distance" value may range from 0 to 25,000 miles. It will be necessary to select a small number of possible values that are significant in the software under test, such as 0, 99, 100, 999, 1000 1001, 99,999 miles. If this selection of values misses some significant point in the code, the test may not detect an error.

Q: The size of a covering array grows exponentially in proportion to what variables? Logarithmically in proportion to what variable? What is the significance of these facts?

A: The array size is proportional to v^t, where v is the number of values per parameter and t is the interaction strength, that is, value of t for t-way testing. However, the number of tests increases only logarithmically with the number of parameters. As a result, the number of values per parameter is a limiting factor in combinatorial testing, but a large number of parameters will not generally be a problem.

Q: What are three ways of handling the oracle problem?

A: Crash testing, embedded assertions, and model-based test generation.

Q: For what value of t must a t-way covering array be generated to catch the following fault?

```
if (A<10 && B>0) {bad code}
else             {good code}
```

A: A 2-way array is needed, so that at least one test contains a combination of values for A and B that will trigger the branch to the bad code, assuming that appropriate equivalence classes have been established.

Q: For what value of *t* must a *t*-way covering array be generated to catch the following fault?

```
if ((A<10 || B>0) && C>90) {bad code}
else                       {good code}
```

A: A 2-way array is needed, because either `A && C` or `B && C` will cause a branch into the bad code.

Q: For what value of *t* must a *t*-way covering array be generated to catch the following fault?

```
if (A<10 && B>0 && C>90)    {bad code}
else                        {good code}
```

A: A 3-way array is needed.

Q: Suppose you are asked to test a program with 10 input parameters, with 100 values each, using a 6-way covering array. Can you generate a covering array for this problem?

A: In theory yes, but not as a practical matter, because the size of a *t*-way covering array is proportional to v^t for parameters with v values, so in this case the number of tests would be proportional to $100^6 = 10^{12}$. The way to approach this problem is to do input space partitioning so that roughly 10 or fewer representative values are used for each parameter. This method is discussed in more detail in later chapters.

Q: If all variables have the same number of values, the number of *t*-way settings of n variables with v values each is $v^t \times C(n, t)$. But in most cases, variables have different numbers of values. For the following five variables with values shown, how many 2-way settings (value pairs) are there: *a*: 0,1,2; *b*: 0,1; *c*: 0,1,2; *d*: 0,1; *e*: 0,1?

A: $6 + 9 + 6 + 6 + 6 + 4 + 4 + 6 + 6 + 4 = 57$. This is $v_a v_b + v_a v_c + v_a v_d + v_a v_e + v_b v_c + v_b v_d + v_b v_e + v_c v_d + v_c v_e + v_d v_e$, where v_i = number of values for variable *i*.

CHAPTER 2

Combinatorial Testing Applied

In Chapter 1, we introduced the interaction rule and noted that if all failures can be triggered by t-way combinations or less, then testing all t-way combinations is in some sense equivalent to exhaustive testing. We have referred to this approach as *pseudoexhaustive* testing [103]: clearly, it is not exhaustive because we do not cover all possible inputs, and issues such as timing and concurrency may be missed. As always, testing is vital but it cannot guarantee the complete absence of errors [36]. But the ability to test all combinations of (discretized) values, up to the interaction strength actually involved in failures, provides a powerful assurance method. Can this approach really detect as many faults as exhaustive testing across discretized values? Although it will take years of experience with combinatorial testing to give a full answer, this chapter summarizes two studies of practical problems where combinatorial testing did indeed find faults as well as exhaustive testing with discretized inputs, using only a fraction of the tests.

2.1 DOCUMENT OBJECT MODEL

Carmelo Montanez-Rivera, D. Richard Kuhn, Mary Brady, Rick Rivello, Jenise Reyes Rodriguez, and Michael Powers

The document object model (DOM) [220] is a standardized method for representing and interacting with the components of XML, HTML, and XHTML documents. DOM lets programs and scripts access and update the content, structure, and style of documents dynamically, making it easier to produce web applications in which pages are accessed nonsequentially. DOM is standardized by the World Wide Web Consortium (W3C). Since its origination in 1996 as a convention for accessing and modifying the parts of Javascript web pages (now known as DOM level 0), DOM has evolved as a series of standards offering progressively greater capabilities. Level 1 introduced a model that allowed changing any part of the HTML document, and level 2 added support for XML namespaces, load and save, cascading stylesheets (CSS), traversing the document, and working with the ranges of content. Level 3 brings additional features, including keyboard event handling.

DOM level 3 events [220] is a W3C working written by the Web Applications Working group. Implemented in browsers, it is a generic platform and language-neutral event system that allows the registration of event handlers, describes event flow through a tree structure, and provides the basic contextual information for each event. This work builds on the previous DOM level 2 event specifications. There were two basic goals in the design of DOM level 3 events. The first goal was to design an event system that allows the registration of event listeners and describes an event flow through a tree structure. The second goal was to provide a common subset of the current event system used on DOM level 3 event browsers.

2.1.1 Constructing Tests for DOM Events

Full exhaustive testing (of discretized inputs) was initially used for DOM conformance, more than 36,000 tests.

DOM browser implementations typically contain tens of thousands of source lines of code. To help ensure successful implementations of this complex standard, National Institute of Standards and Technology (NIST) developed the DOM conformance test suites [221], which include tests for many DOM components. Early DOM tests were handcoded in a test language, and then processed to produce ECMAScript and Java. In the current version of the test suites, the tests are specified in an XML grammar, allowing easy mapping from specification to a variety of language bindings. Since the grammar is generated automatically [142] from the DOM specs, tests can be constructed quickly and correctly. The output of the test generation process includes the following components, which implementers can use in testing their product for DOM interoperability:

- Tests in the XML representation language
- XSLT stylesheets necessary to generate the Java and ECMAScript bindings
- Generated executable code

Tests for 35 (out of 36) DOM events were generated. The specification defines each event with an interface definition language (IDL), which in turn defines a number of functions for each event. A typical function can have anywhere from one to 15 parameters. The application programming interface (API) for each function is defined as an XML infoset, which specifies the abstract data model of the XML document using a predefined set of tags. The XML infosets were programmatically generated through a Java application. Since the IDL definition could be accessed directly from the specs website, the web address was given as an input to the Java application. This way, the application could read and traverse them, extracting just the information of interest. In this case, the function names and their respective parameters, argument names, and so on became part of the XML file that was used to feed a test generation tool to automatically create the DOM level 3 tests.

The conventional category partitioning was used to select the representative values for non-Boolean parameters. The initial test set was exhaustive across the equivalence classes, producing 36,626 tests that exercised all possible combinations of equivalence class elements. Note that this is not fully exhaustive—all possible value combinations—because such a test suite is generally intractable with continuous-valued parameters. However, it is exhaustive with respect to the equivalence class elements. Thus, this test suite will be referred to as the *exhaustive test suite* in the remainder of this section. Two different implementations were tested. The DOM events and the number of tests for each implementation are shown in Table 2.1. This set of exhaustive tests detected a total of 72 failures. The automated tools made it possible to construct tests programmatically, greatly reducing the effort required for testing. However, human intervention is required to run individual tests and evaluate test results; so, the conformance testing team sought ways to reduce the number of tests required without sacrificing quality.

TABLE 2.1 DOM Level 3 Event Tests: Exhaustive

Event Name	Parameter	Tests
Abort	3	12
Blur	5	24
Click	15	4352
Change	3	12
dblClick	15	4352
DOMActivate	5	24
DOMAttrModified	8	16
DOMCharacterDataModified	8	64
DOMElementNameChanged	6	8
DOMFocusIn	5	24
DOMFocusOut	5	24
DOMNodeInserted	8	128
DOMNodeInsertedIntoDocument	8	128
DOMNodeRemoved	8	128
DOMNodeRemovedFromDocument	8	128
DOMSubTreeModified	8	64
Error	3	12
Focus	5	24
KeyDown	1	17
KeyUp	1	17
Load	3	24
MouseDown	15	4352
MouseMove	15	4352
MouseOut	15	4352
MouseOver	15	4352
MouseUp	15	4352
MouseWheel	14	1024
Reset	3	12
Resize	5	48
Scroll	5	48
Select	3	12
Submit	3	12
TextInput	5	8
Unload	3	24
Wheel	15	4096
Total tests		36,626

2.1.2 Combinatorial Testing Approach

Since the DOM test suite had already been applied with exhaustive (across equivalence values) tests against a variety of implementations, it provided a valuable opportunity to evaluate combinatorial testing on real-world software. If results showed that a much smaller test suite could achieve the same level of fault detection, then testing could be done at much less cost in staff time and resources. An obvious critical question in using this approach is—*what level of* t-*way interaction is necessary*? Can all faults be detected with two-way (pairwise) tests, or does the application require three-way, four-way, or higher-strength tests? This work helped to address these questions as well.

To investigate the effectiveness of combinatorial testing, covering arrays of two-way through six-way tests were produced, using ACTS. Using t-way combinations can significantly reduce the number of tests as compared with exhaustive tests. For example, the `mousedown` event (Figure 2.1) requires 4352 tests if all combinations are to be realized. Combinatorial testing reduces the set to 86 tests for four-way coverage.

Table 2.2 details the number of parameters and the number of tests produced for each of the 35 DOM events, for $t = 2$ through 6. That is, the tests covered all two-way through six-way combinations of values. Note that for events with few parameters, the number of tests is the same for the original test suite (Table 2.1) and is combinatorial for various levels of t. For example, 12 tests were produced for Abort in the original testing and also for combinatorial testing at $t = 3$ through 6. This is because producing all n-way combinations for n variables is simply all possible combinations of these n variables, and Abort has three variables. This situation is not unusual when testing configurations with a limited number of values for each parameter. For nine of the 35 events (two Click events, six Mouse events, and Wheel), all combinations are not covered even with six-way tests. For these events, combinatorial testing provides a significant gain in efficiency.

2.1.3 Test Results

Four-way combinatorial testing found all DOM faults using 95% fewer tests than the original test suite.

Table 2.3 shows the faults detected for each event. All conditions flagged by the exhaustive test suite were also detected by three of the combinatorial

```
<event name = "mousedown" interface = "MouseEvent">
  <targetNodes>
     <value>Element</value>
 </targetNodes>
<function name = "initMouseEvent" level = "2">
 <eargument name = "typeArg" type = "DOMString" targetAttribute = "type">
  <eargValue value = "mousedown"/>
 </eargument>
 <eargument name = "canBubbleArg" type = "boolean" targetAttribute = "bubbles'
   <eargValue value =  "true"/>
   <eargValue value =  "false"/>
 </eargument>
 <eargument name = "cancelableArg" type = "boolean" targetAttribute = "cancel
   <eargValue value =  "true"/>
   <eargValue value =  "false"/>
 </eargument>
 <eargument name = "viewArg" type = "views::AbstractView" targetAttribute = "
   <eargValue value = "window"/>
 </eargument>
 <eargument name = "detailArg" type = "long" targetAttribute = "detail">
   <eargValue value =  "5"/>
   <eargValue value =  "-5"/>
 </eargument>
 <eargument name = "screenXArg" type = "long" targetAttribute = "screenX">
   <eargValue value =  "5"/>
   <eargValue value =  "-5"/>
 </eargument>
 <eargument name = "screenYArg" type = "long" targetAttribute = "screenY">
   <eargValue value =  "5"/>
   <eargValue value =  "-5"/>
 </eargument>
 <eargument name = "clientXArg" type = "long" targetAttribute = "clientX">
   <eargValue value =  "5"/>
   <eargValue value =  "-5"/>
 </eargument>
 <eargument name = "clientYArg" type = "long" targetAttribute = "clientY">
   <eargValue value =  "5"/>
   <eargValue value =  "-5"/>
 </eargument>
 <eargument name = "ctrlKeyArg" type = "boolean" targetAttribute = "ctrlKey">
   <eargValue value =  "true"/>
   <eargValue value =  "false"/>
 </eargument>
 <eargument name = "altKeyArg" type = "boolean" targetAttribute = "altKey">
   <eargValue value =  "true"/>
   <eargValue value =  "false"/>
 </eargument>
 <eargument name = "shiftKeyArg" type = "boolean" targetAttribute = "shiftKey'
   <eargValue value =  "true"/>
   <eargValue value =  "false"/>
 </eargument>
 <eargument name = "metaKeyArg" type = "boolean" targetAttribute = "metaKey">
   <eargValue value =  "true"/>
   <eargValue value =  "false"/>
 </eargument>

      ...

 </function>
</event>
```

FIGURE 2.1 XML infosets generated from IDLs used as an input to the test accelerator.

TABLE 2.2 DOM Three-Level Tests: Combinatorial

Event Name	Number of Parameters	Two-Way Tests	Three-Way Tests	Four-Way Tests	Five-Way Tests	Six-Way Tests
Abort	3	8	12	12	12	12
Blur	5	10	16	24	24	24
Click	15	18	40	86	188	353
Change	3	8	12	12	12	12
dblClick	15	18	40	86	188	353
DOMActivate	5	10	16	24	24	24
DOMAttrModified	8	8	16	16	16	16
DOMCharacterDataModified	8	32	62	64	64	64
DOMElementNameChanged	6	8	8	8	8	8
DOMFocusIn	5	10	16	24	24	24
DOMFocusOut	5	10	16	24	24	24
DOMNodeInserted	8	64	128	128	128	128
DOMNodeInsertedIntoDocument	8	64	128	128	128	128
DOMNodeRemoved	8	64	128	128	128	128
DOMNodeRemovedFromDocument	8	64	128	128	128	128
DOMSubTreeModified	8	32	64	64	64	64
Error	3	8	12	12	12	12
Focus	5	10	16	24	24	24
KeyDown	1	9	17	17	17	17
KeyUp	1	9	17	17	17	17
Load	3	16	24	24	24	24
MouseDown	15	18	40	86	188	353
MouseMove	15	18	40	86	188	353
MouseOut	15	18	40	86	188	353
MouseOver	15	18	40	86	188	353
MouseUp	15	18	40	86	188	353
MouseWheel	14	16	40	82	170	308
Reset	3	8	12	12	12	12
Resize	5	20	32	48	48	48
Scroll	5	20	32	48	48	48
Select	3	8	12	12	12	12
Submit	3	8	12	12	12	12
TextInput	5	8	8	8	8	8
Unload	3	16	12	24	24	24
Wheel	15	20	44	92	214	406
Total tests		702	1342	1818	2742	4227

TABLE 2.3 Comparison of t-Way with Exhaustive Test Set Size

			Test Results		
t	Tests	Percentage of Original	Pass	Fail	Not Run
2	702	1.92	202	27	473
3	1342	3.67	786	27	529
4	1818	4.96	437	72	1309
5	2742	7.49	908	72	1762
6	4227	11.54	1803	72	2352

testing scenarios (four-, five-, and six-way testing), which indicates that the implementation faults were triggered by four-way interactions or less. Pairwise testing would have been inadequate for the DOM implementations because two-way and three-way tests detected only 37.5% of the faults. As can be seen in Table 2.3, the exhaustive (all possible combinations) tests of equivalence class elements and the four-way through six-way combinatorial tests were equally successful in fault detection, indicating that exhaustive testing added no benefit. These findings are consistent with earlier studies [105] that showed that software faults are triggered by interactions of a small number of variables, for applications in a variety of domains. DOM testing was somewhat unusual in that exhaustive testing of equivalence class elements was possible at all. For most software, too many possible input combinations exist to cover even a tiny fraction of the exhaustive set; so, combinatorial methods may be of even greater benefit for these.

The original test suite contained a total of 36,626 tests (Table 2.1) for all combinations of events, but after applying combinatorial testing, the set of tests is dramatically reduced depending on the level of t-way interactions tested, as shown in Table 2.3. Table 2.3 also shows the results for two-way through six-way testing. Note that although the number of tests that successfully execute varies among t-way combination, the number of failures remains a constant at $t = 2$ and 3, and at $t = 4$ through 6. The last column shows the tests that did not execute to completion, in almost all cases due to nonsupport of the feature under test.

DOM results were consistent with previous findings that the number of parameters interacting in failures is small (in this case, four-way). Comparing the results of the DOM testing with previous data on t-way interaction failures (Figure 2.2), we can see that some DOM failures were more difficult to detect, in the sense that a smaller percentage of the total DOM failures was found by two-way and three-way tests than for the other

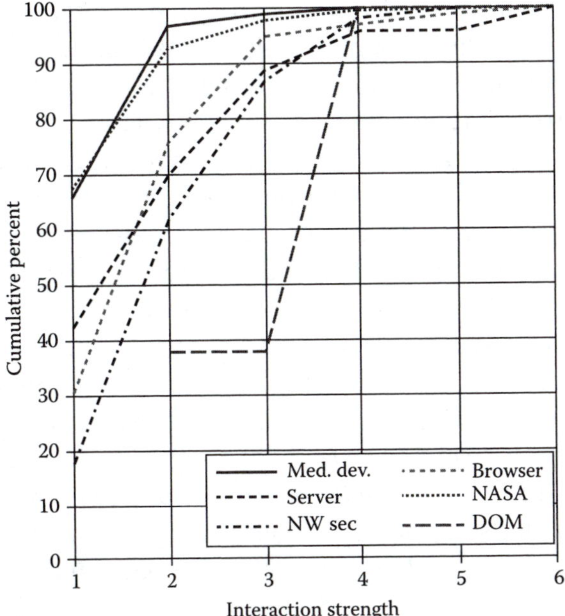

FIGURE 2.2 **(See color insert.)** DOM compared with other applications.

application domains. The unusual shape of the curve for DOM tests may result from the large number of parameters for which exhaustive coverage was reached (so that the number of tests remained constant after a certain point). Thus, there are two sets of events: a large set of 26 parameters with a few possible values that could be covered exhaustively with two-way or three-way tests, and a smaller set with a larger input space (from 1024 to 4352). In particular, nine events (`click`, `dblClick`, `mouse` events, and `wheel`) have approximately the same input space size, with the number of tests increasing at the same rate for each event, while for the rest of the events, exhaustive coverage is reached at either $t = 2$ or $t = 3$. The ability to compare results of the previously conducted exhaustive testing with combinatorial testing provides an added measure of confidence in the applicability of these methods to this type of interoperability testing.

2.1.4 Cost and Practical Considerations

The DOM events testing shows that combinatorial testing can significantly reduce the cost and time required for conformance testing for standards with characteristics similar to DOM. What is the appropriate interaction strength to use in this type of testing? Intuitively, it seems that if no additional faults are detected by t-way tests, then it may be reasonable to conduct additional testing only for $t + 1$ interactions but it may not be greater if no additional faults are found at $t + 1$. In empirical studies of software failures, the number of faults detected at $t > 2$ decreased monotonically with t, and the DOM testing results are consistent with this earlier finding. Following this strategy for the DOM testing would result in running two-way tests through five-way, then stopping because no additional faults were detected beyond the four-way testing. Alternatively, given the apparent insufficient fault detection of pairwise testing, testers may prefer to standardize on a four-way or a higher level of interaction coverage. This option may be particularly attractive for an organization that produces a series of similar products and has enough experience to identify the most cost-effective level of testing. Even the relatively strong four-way testing in this example was only 5% of the original test set size. The results of this study have been sufficiently promising for combinatorial methods to be applied in testing other interoperability standards.

2.2 RICH WEB APPLICATIONS

Chad M. Maughan

Rich web applications (RWA) are rapidly gaining in popularity. RWAs focus on creating a rich, interactive experience that is browser independent and does not rely on proprietary, client-side browser plug-ins such as Adobe Flash. Typically, an RWA consists of a single HTML page that acts as an entry point. AJAX calls to a stateless (i.e., sessionless) server retrieve data that is then used to dynamically manipulate the DOM. Semantic (i.e., human understandable) URL fragments control history and browser interactions. Client-side JavaScript frameworks such as Google Web Toolkit (GWT), Backbone.js, Angular.js, and Ember.js facilitate the movement of logic and processing from the server side to the client side in the browser while helping to deliver a rich user experience.

While end users may benefit from the migration of complex application logic from the server to the browser, new testing challenges arise. The exceptions and coding irregularities no longer occur solely on the server in a convenient, controlled, and centralized location. Errors now occur in a distributed manner in various types and versions of browsers, computers, and environments in a dynamically typed language that is difficult to statically analyze. Maughan [129] proposes the use of combinatorial testing for both server and client sides.

The client-side migration of logic poses several problems. Maughan addresses two specific problems [129]. First, semantic URL structures contain no formal format standard similar to RFC 1738 [32] for traditional URL formats. This makes the systematic discovery of independent variables difficult for combinatorial test generation in RWA URLs. Maughan presents an approach for identifying variables in semantic URLs using the analysis of hierarchical branching complexity of URL structures. Second, JavaScript exceptions are difficult to systematically identify and report. Maughan also demonstrates, using the available frameworks and libraries, how to systematically capture JavaScript exceptions directly from the browser.

2.2.1 Systematic Variable Detection in Semantic URLs

Wang et al. propose building a navigation graph [208] to identify "abstract URLs" that follow RFC 1738 [32,80]. A similar navigation graph can be made for a web application employing semantic URLs. For example, a semantic URL, such as http://example.com/#/state/utah/county/cache/city/logan contains the variables state, county, and city that a tester may extract. Carrying out a depth-first traversal of a web application to build a navigation graph gathers all the available URLs. To extract the variables in an RWA with semantic URLs, Maughan proposes an algorithm that uses graph theory and branching complexity to identify variables [129].

Working with the sample URL listed above, the following steps (Figure 2.3) would be taken, lopping off the outer portions of the URL fragment and calculating the branching complexity on each vertex for later analysis.

- http://example.com/#/state/utah/county/cache/city/logan
- http://example.com/#/state/utah/county/cache/city
- ... (additional steps omitted for brevity)
- http://example.com/#/state
- http://example.com/#

2.2.2 JavaScript Fault Classification and Identification

Another challenge presented in RWAs is capturing JavaScript errors. While developers may employ browser plug-ins such as Firebug for Firefox or the Chrome Developer Tools to see exceptions by enabling the

```
foreach (url in urls)
        Node previousNode = retrieveOrCreateNode(url);
        int position = url.length();
        String shortenedUrl;
        while (position > fragmentIdentifier) {
                position = url.lastIndexOf("/");
                if (position > 0) {
                url = url.substring(0, position);
                        Node node = retrieveOrCreateNode(url);

                        // increments branch complexity
                        createRelationship(previousNode, node);

                        // determine variance of branch 'similarities'
                        calculateSubBranchSimilarities(node);

                        previousNode = node;
                }
                else {
                        break;
                }
        }
        Node last = retrieveOrCreateNode(shortenedUrl);
}
```

FIGURE 2.3 URL processing algorithm.

JavaScript console, the end users are typically unaware of these errors. We propose using the available tools and libraries for capturing these errors.

To capture these exceptions, Maughan uses the Selenium WebDriver, BrowserMob Proxy Server, and navigation graph described earlier to intercept HTTP responses from the server to the Selenium client. He then injects the HTML <HEAD> element with custom JavaScript that stores exceptions as Selenium interacts with the application, then reports back those JavaScript errors. Additionally, the embedded proxy server also records network exceptions or any non-200 HTTP status code responses received for the missing resources.

Guo et al. [80] expanded on Sampath et al.'s [170,171] web application fault classification by adding seven subcategories to logic faults. Not all of Guo's faults apply to RWA. Ocariza [148] observed that 94% of JavaScript exceptions generally fall into five categories: permission-denied errors, null exception errors, undefined symbol errors, and miscellaneous. We use these categories to seed faults in the empirical study in the next section.

2.2.3 Empirical Study

To facilitate an empirical study on the effectiveness of combinatorial testing on an RWA, Maughan constructed a mobile-optimized sample application and seeded it with 77 faults from the classifications of the previous section. A sample screenshot of the application is shown in Figure 2.4. This application shows the area, density, and other census information for U.S. states, counties, and cities that have a population of over 25,000

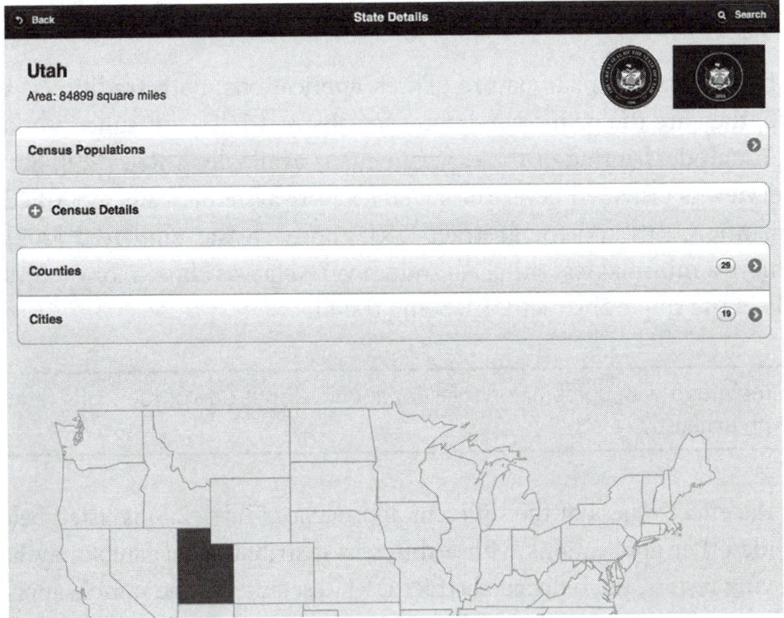

FIGURE 2.4 **(See color insert.)** Screenshot of the sample RWA.

TABLE 2.4 Sample Rich Web Application Metrics

Metrics	
Number of application states	199,484
Number of files	328
JavaScript lines of code	605
Java lines of code	1091
Seeded faults	77

residents. The characteristics of the application are provided in Table 2.4. The application is available at http://chadmaughan.com/thesis. The focus of the study was to answer the following:

- What is the impact on the size of the test suite using abstract URLs instead of an exhaustive enumeration of every variable combination?
- What is the fault-finding effectiveness of testing with abstract URLs versus exhaustive testing?
- What is the impact on the size of the test suite using combinatorial coverage compared to single coverage?
- What is the fault-finding effectiveness of the combinatorial coverage compared with single coverage?

Due to the template nature of web applications, both traditional and rich, they are the prime candidates for abstract URL test suites. A more thorough description of the experiments is available in Ref. [129], but an overview is provided here. The variables were systematically identified in semantic URLs and the abstract URL approach was employed [208] to achieve a minimal test suite. A random variable was chosen for each variable in the single abstract URL being tested.

Simple two-way combinatorial testing found all but one of the faults found with exhaustive tests.

The effectiveness of the different approaches (Table 2.5) is listed below. While other applications with a different distribution of faults may have varying results, the reduced abstract URL test suite on the sample application found the fewest faults, followed by combinatorial test suite, and then exhaustive test suite. The abstract URL test suite identified all the JavaScript

TABLE 2.5 Strategy Test Counts

Strategy	Number of Tests
Single coverage (exhaustive)	199,484
Two-way combinatorial coverage	25,560
Abstract URLs	27

TABLE 2.6 Strategy Success Rate

Strategy	Faults Found	Percentage
Single coverage (exhaustive)	69	89.6
Two-way combinatorial	68	88.3
Abstract URLs	15	19.5

exceptions but missed nearly all data-related faults. The abstract URL effectiveness may vary depending on what random tiebreaking value is selected for the test case. Additionally, the single coverage (i.e., exhaustive) testing did not discover all faults due to a single browser (Firefox in this experiment) being used. Other browsers raised permission-denied errors (i.e., violations of the browser's same-origin policy) whereas Firefox did not. It is striking that the two-way combinatorial test suite was significantly smaller, yet it found only one less fault (that would have been found by three-way tests) than the single coverage (exhaustive) test suite (Table 2.6).

2.3 CHAPTER SUMMARY

The realistic case studies introduced in this chapter help to validate the interaction rule and illustrate how combinatorial testing can be applied in practical applications. DOM conformance testing applied combinatorial methods to drastically reduce the number of tests required, finding all faults with <5% of the original (exhaustive) test suite. This reduction was particularly significant because the DOM standard is large and complex, and testing its embedded code (tens of thousands of lines) in browsers is time consuming. Consistent with the empirical studies reported earlier, all faults detected by exhaustive testing were found with a four-way array of tests.

Combinatorial testing also shows promise for testing RWAs in the browser, using an automated approach to generate test suites based on single coverage (exhaustive), two-way combinatorial, and abstract URLs. The results are striking that the two-way combinatorial approach used a test suite that was only ~12.8% of the exhaustive test suite, but found only one fewer fault than the exhaustive approach. This motivates future work to examine the application of our approach on additional RWAs. The complexity of RWAs is growing with the demand for sophisticated web applications; so, this method can dramatically reduce the cost of testing for such products.

CHAPTER 3

Configuration Testing

THE TERM "CONFIGURATION" MAY be used in slightly different ways with respect to software. In some cases, it may refer to options that are settable through an external file or other source. For example, a database management system may have configurable options for storage location and size, maximum size of various tables, key length, and other aspects of databases. These configurable options are read in when the system is initialized and used to set properties of the application. In other cases, configuration refers to characteristics of the platform on which the application is running, such as the presence or absence of a hard keyboard on a smart phone, the network protocol used, or the type of database. In this case, the configurable options are expected to essentially provide the same functions to the software—network interface or searchable storage—but low-level functions in the application must interface differently depending on the protocol or the database in use. The software is built to operate correctly on a variety of platforms, and different parts of the code may be exercised depending on the configuration.

3.1 RUNTIME ENVIRONMENT CONFIGURATIONS

One of the most common problems in software testing is assuring that an application can run on a variety of platforms. Different operating systems, web browsers, network protocols, or databases may be operated by customers, but developers would like to ensure that their software runs correctly on all platforms. An example illustrating the complexity of the problem occurred in July, 2012. A major antivirus program suffered crashes on certain configurations of Windows XP machines. According to a *Register* news article [120],

> Subsequent analysis has revealed that a three-way clash between third-party encryption drivers, Symantec's own security software and the Windows XP Cache manager resulted in the infamous Blue Screen of Death (BSOD) on vulnerable machines, as this advisory explains:
>
> > The root cause of the issue was an incompatibility due to a three-way interaction between some third-party software that implements a file system driver using kernel stack based file objects—typical of encryption drivers, the SONAR signature and the Windows XP Cache manager. The SONAR signature update caused new file operations that create the conflict and led to the system crash.

Combinatorial testing of runtime configurations can help in catching this type of problem. While it is rarely practical to test all possible runtime platforms, the methods described in this chapter can be used for efficient testing of all t-way combinations of platform configurations.

Returning to the simple example (Table 3.1) introduced in Chapter 1, we illustrate the development of test configurations and compare the size of test suites for various interaction strengths versus testing all possible configurations. For the five configuration parameters, we have $3 \times 2 \times 2 \times 2 \times 3 = 72$ configurations. Note that at $t = 5$, the number of tests is the same as exhaustive testing for this example because there are only five parameters. The savings as a percentage of exhaustive testing are good, but they are not that impressive for this small example. With larger systems, the savings can be enormous, as will be seen in the next section.

After the parameters and possible values for each test suite have been determined, a covering array can be generated using a software tool. In this chapter, the generation process will be illustrated using the ACTS covering array tool, which is described in more detail in Appendix D, but other tools

TABLE 3.1 Simple Example Configuration Options

Parameter	Values
Operating system	XP, OS X, RHL
Browser	IE, Firefox
Protocol	IPv4, IPv6
CPU	Intel, AMD
DBMS	MySQL, Sybase, Oracle

```
[System]

[Parameter]
OS (enum): XP,OS_X,RHL
Browser (enum): IE, Firefox
Protocol(enum): IPv4,IPv6
CPU (enum):  Intel,AMD
DBMS (enum): MySQL,Sybase,Oracle

[Relation]
[Constraint]

[Misc]
```

FIGURE 3.1 ACTS input includes the names of parameters, types, and possible values.

may have similar features. In addition to the summary in Appendix D, a comprehensive user manual is included with the ACTS download.

The first step in creating test configurations is to specify the parameters and possible values, as shown in Figure 3.1. Another covering array tool or the GUI version of ACTS would of course have a different specification, but the essential features will be similar to Figure 3.1.

The degree of interaction must also be specified: two-way, three-way, coverage, and so on. The output can be created as a matrix of numbers, comma-separated value, or Excel spreadsheet format. If the output will be used by human testers rather than as an input for further machine processing, the format in Figure 3.2 is useful.

The test set size for two-way combinations is shown in Table 3.2. Only 10 tests are needed. Moving to three-way or higher-interaction strengths requires more tests, as shown in Table 3.2.

In this example, substantial savings could be realized by testing t-way configurations instead of all possible configurations, although for some applications (such as a small but highly critical module), a full exhaustive test may be warranted. As we will see in the next example, in many cases, it is impossible to test all configurations; so, we need to develop reasonable alternatives.

3.2 HIGHLY CONFIGURABLE SYSTEMS AND SOFTWARE PRODUCT LINES

A software product line with n features may produce 2^n products.

```
Degree of interaction coverage: 2
Number of parameters: 5
Maximum number of values per parameter: 3
Number of configurations: 10
-------------------------------------
Configuration #1:

1 = OS=XP
2 = Browser=IE
3 = Protocol=IPv4
4 = CPU=Intel
5 = DBMS=MySQL
-------------------------------------
Configuration #2:

1 = OS=XP
2 = Browser=Firefox
3 = Protocol=IPv6
4 = CPU=AMD
5 = DBMS=Sybase
-------------------------------------
Configuration #3:

1 = OS=XP
2 = Browser=IE
3 = Protocol=IPv6
4 = CPU=Intel
5 = DBMS=Oracle
-------------------------------------
Configuration #4:

1 = OS=OS_X
2 = Browser=Firefox
3 = Protocol=IPv4
4 = CPU=AMD
5 = DBMS=MySQL
etc .
```

FIGURE 3.2 Excerpt of the test configuration output covering all two-way combinations.

TABLE 3.2 Number of Combinatorial Tests for a Simple Example

t	**# Tests**	**% of Exhaustive**
2	10	14
3	18	25
4	36	50
5	72	100

Software product lines (SPLs) are an increasingly attractive approach to the development of an application. An SPL uses standardized development procedures on systems that "share a common, managed set of features satisfying the specific needs of a particular market segment or mission and that are developed from a common set of core assets" [187]. The basic idea of an SPL is that enterprises or their subunits tend to produce families of software products for a particular application domain or market [87,145,163]. For example, a company may develop software products for point-of-sale (POS) and retail store management. By combining software that implements various features, a wide variety of products can be provided with far less effort than the traditional approaches of development. In the retail store management example, a basic POS terminal application may allow for input from the cashier's keyboard or may allow for a laser scanner embedded in the checkout counter, whereas a more sophisticated terminal application may add features for a handheld scanner and a scale. Thus, in some cases, a product line can thus be viewed as a framework that can produce 2^n products where there are n different features [87]. With the high degree of customization and configurable feature sets, combinatorial testing can be effective especially when applied to SPLs [58,59,98,152].

Telecommunications and mobile phone vendors have been among the early adopters of the SPL approach, with significant success [186]. Smart phones have become enormously popular because they combine communication capability with powerful graphical displays and processing capability. Literally tens of thousands of smart phone applications, or "apps," are developed annually. Among the platforms for smart phone apps is the Android, which includes an open-source development environment and a specialized operating system. Android units contain a large number of configuration options that control the behavior of the device. Android apps must operate across a variety of hardware and software platforms since not all products support the same options. For example, some smart phones may have a physical keyboard and others may present a soft keyboard using the touch-sensitive screen. Keyboards may also be either only numeric with a few special keys, or a full typewriter keyboard. Depending on the state of the app and user choices, the keyboard may be visible or hidden. Ensuring that a particular app works across the enormous number of options is a significant challenge for developers. The extensive set of options makes it intractable to test all possible configurations; so, combinatorial testing is a practical alternative.

Figure 3.3 shows a resource configuration file for Android apps. A total of 35 options may be set. Our task is to develop a set of test configurations

```
int HARDKEYBOARDHIDDEN_NO;
int HARDKEYBOARDHIDDEN_UNDEFINED;
int HARDKEYBOARDHIDDEN_YES;
int KEYBOARDHIDDEN_NO;
int KEYBOARDHIDDEN_UNDEFINED;
int KEYBOARDHIDDEN_YES;
int KEYBOARD_12KEY;
int KEYBOARD_NOKEYS;
int KEYBOARD_QWERTY;
int KEYBOARD_UNDEFINED;
int NAVIGATIONHIDDEN_NO;
int NAVIGATIONHIDDEN_UNDEFINED;
int NAVIGATIONHIDDEN_YES;
int NAVIGATION_DPAD;
int NAVIGATION_NONAV;
int NAVIGATION_TRACKBALL;
int NAVIGATION_UNDEFINED;
int NAVIGATION_WHEEL;
int ORIENTATION_LANDSCAPE;
int ORIENTATION_PORTRAIT;
int ORIENTATION_SQUARE;
int ORIENTATION_UNDEFINED;
int SCREENLAYOUT_LONG_MASK;
int SCREENLAYOUT_LONG_NO;
int SCREENLAYOUT_LONG_UNDEFINED;
int SCREENLAYOUT_LONG_YES;
int SCREENLAYOUT_SIZE_LARGE;
int SCREENLAYOUT_SIZE_MASK;
int SCREENLAYOUT_SIZE_NORMAL;
int SCREENLAYOUT_SIZE_SMALL;
int SCREENLAYOUT_SIZE_UNDEFINED;
int TOUCHSCREEN_FINGER;
int TOUCHSCREEN_NOTOUCH;
int TOUCHSCREEN_STYLUS;
int TOUCHSCREEN_UNDEFINED;
```

FIGURE 3.3 Android resource configuration file.

that allows testing across all four-way combinations of these options. The first step is to determine the set of parameters and possible values for each option that will be tested. Although the options are listed individually to allow a specific integer value to be associated with each option, they clearly represent sets of option values with mutually exclusive choices. For example, "Keyboard Hidden" may be "yes," "no," or "undefined." These values will be the possible settings for parameter names that we will use in generating a covering array. Table 3.3 shows the names of parameters and the

TABLE 3.3 This Set of Android Options Has 172,800 Possible Configurations

Parameter Name	Values	# Values
HARDKEYBOARDHIDDEN	NO, UNDEFINED, YES	3
KEYBOARDHIDDEN	NO, UNDEFINED, YES	3
KEYBOARD	12KEY, NOKEYS, QWERTY, UNDEFINED	4
NAVIGATIONHIDDEN	NO, UNDEFINED, YES	3
NAVIGATION	DPAD, NONAV, TRACKBALL, UNDEFINED, WHEEL	5
ORIENTATION	LANDSCAPE, PORTRAIT, SQUARE, UNDEFINED	4
SCREENLAYOUT_LONG	MASK, NO, UNDEFINED, YES	4
SCREENLAYOUT_SIZE	LARGE, MASK, NORMAL, SMALL, UNDEFINED	5
TOUCHSCREEN	FINGER, NOTOUCH, STYLUS, UNDEFINED	4

number of possible values that we will use for input to the covering array generator. For a complete specification of these parameters, see http://developer.android.com/reference/android/content/res/Configuration.html

Using Table 3.3, we can calculate the total number of configurations: $3 \times 3 \times 4 \times 3 \times 5 \times 4 \times 4 \times 5 \times 4 = 172{,}800$ configurations (i.e., a $3^3 4^4 5^2$ system). Similar to many applications, thorough testing will require some human intervention to run tests and to verify results, and a test suite will typically include many tests. If each test suite can be run in 15 min, it will roughly take 24 staff-years to complete the testing for an app. With salary and benefit costs for each tester of $150,000, the cost of testing an app will be more than $3 million, making it virtually impossible to return a profit for most apps. We saw in Section 3.1 that combinatorial methods can reduce the number of tests needed for strong assurance, but will the reduction in test set size be enough to provide effective testing for apps at a reasonable cost?

Using the covering array tool, we can produce tests that cover *t*-way combinations of values. Table 3.4 shows the number of tests required at several levels of *t*. For many applications, 2-way or 3-way testing may be appropriate, and either of these will require <1% of the time required to cover all possible test configurations. This example illustrates the power of combinatorial testing for real-world testing, and how its advantages increase with the size of the problem.

TABLE 3.4 Number of Combinatorial Tests Is a Fraction of an Exhaustive Test Set

t	# Tests	% of Exhaustive
2	29	0.02
3	137	0.08
4	625	0.4
5	2532	1.5
6	9168	5.3

3.3 INVALID COMBINATIONS AND CONSTRAINTS

So far, we have assumed that the set of possible values for parameters never changes. Thus, a covering array of *t*-way combinations of possible values would contain combinations that either would occur in the systems under test, or could occur and must therefore be tested. But look more closely at the configurations in Table 1.1 of Chapter 1. In practice, the Internet Explorer browser is never used on Linux systems; so, it would be impossible to create a configuration that specified IE on a Linux system. This is an example of a constraint between the possible values of parameters. Some combinations never occur in practice, or occur only sometimes. Practical testing requires the consideration of constraints.

3.3.1 Constraints among Parameter Values

The system described earlier illustrates a common situation in all types of testing: some combinations cannot be tested because they do not exist for the systems under test. In this case, if the operating system is either OS X or Linux, Internet Explorer is probably not present. Note that we cannot simply delete tests with these untestable combinations because that would result in losing other combinations that are essential to test but are not covered by other tests. For example, deleting tests 5 and 7 in Table 1.1 of Chapter 1 would mean that we would also lose the test for Linux with the IPv6 protocol.

Some combinations never occur in practice.

One way around this problem is to delete tests and to supplement the test suite with manually constructed test configurations to cover the deleted combinations, but covering array tools offer a better solution. With ACTS, we can specify constraints, which tell the tool not to include specified combinations in the generated test configurations. ACTS supports a set of commonly used logic and arithmetic operators to specify the constraints. In this case, the following constraint can be used to ensure that invalid combinations are not generated. It says that if the OS is not XP, then the Browser will be Firefox

```
(OS !="XP") => (Browser="Firefox")
```

The covering array tool will then generate a set of test configurations that does not include the invalid combinations, but it does cover all those test configurations that are essential. The revised test configuration array is shown in Table 3.5. The parameter values that have changed from the original configurations are underlined. Note that adding the constraint also resulted in reducing the number of test configurations by one. This will not always be the case, depending on the constraints used, but it illustrates how constraints can sometimes reduce the problem. Even if particular combinations are testable, the test team may consider some combinations unnecessary (e.g., if it is known that they cannot interact), and constraints could be used to prevent these combinations, possibly reducing the number of test configurations.

In many practical cases, the situation will not be quite as simple as the example above. For example, instead of dealing with only one Windows OS variety (in this case XP), we may have several varieties: XP, Vista, Win7, and Win8. Similarly, there may be many Linux releases to consider, such as Red Hat, Ubuntu, Fedora, and many others plus different releases of the individual Linux versions. Such a situation could lead to very complicated constraint expressions. One approach proposed for handling this problem

TABLE 3.5 Test Configurations for Simple Example with Constraint

Test	OS	Browser	Protocol	CPU	DBMS
1	XP	IE	IPv4	Intel	MySQL
2	XP	Firefox	IPv6	AMD	Sybase
3	XP	IE	IPv6	Intel	Oracle
4	OS X	Firefox	IPv4	AMD	MySQL
5	OS X	Firefox	IPv4	Intel	Sybase
6	OS X	Firefox	IPv6	AMD	Oracle
7	RHL	Firefox	IPv6	Intel	MySQL
8	RHL	Firefox	IPv4	Intel	Oracle
9	XP	IE	IPv4	AMD	Sybase

is the notion of *properties* [177], which can be used to combine related values. For the example here, there could be an "OSfamily" property defined for the OS parameter; so, the constraint could be expressed as

```
(OS.OSfamily !="Windows") => (Browser="Firefox")
```

Without the properties feature, we would need to write something such as

```
(OS !="XP" && OS !="Vista" && OS !="Win7" && OS !="Win8")
      => (Browser="Firefox")
```

If we needed other constraints to also include references to the OSfamily property, the constraint set could become complicated very quickly. Such situations are not uncommon in practical testing.

Although the "properties" feature is not available on most covering array generators, we can achieve the goal of simplifying constraint expression in a different (though somewhat less elegant) way by taking advantage of the power of constraint solvers in ACTS or other tools, along with a little textual substitution. For example, define a global term "WindowsVersion" as

```
(OS="XP" || OS="Vista" || OS="Win7" || OS="Win8")
```

Then constraints can be written such as `!WindowsVersion => (Browser = "Firefox")`. Substituting the parenthetical expression above for "WindowsVersion" using a preprocessor, or simply a text editor will then introduce the necessary expression throughout the constraint set.

3.3.2 Constraints among Parameters

A second way in which untestable combinations may arise in practice is where some parameters become inactive when others are set to particular values. In the previous section, we considered situations where particular parameter values do not occur in combination with other particular values, but the parameters themselves were always present. For example, every test configuration included both the operating system and browser, even though certain OS/browser value combinations did not occur. But for some test problems, a value in one parameter affects not just the possible values for another parameter, but the presence of other parameters themselves, regardless of the values. Returning to the testing problem described in Section 3.1, suppose testers also wanted to consider additional software that may be present in configurations. Java and Microsoft .Net are used by many applications, and it is important to test for compatibility with different versions of these platforms. Thus, it may be desirable to add two additional parameters: "java_version" and "dot_net_version." However, Java can be present on both Windows and Linux platforms, but we must deal with the problem that .Net will not be present on a Linux system. This restriction cannot be handled with an ordinary constraint because if the platform is Linux, the "*dot_net_version*" parameter does not make any sense. Instead, we end up with two different parameter sets: for Windows, the parameters are *OS, browser, protocol, cpu, dbms, java_version, and dot_net_version*; for Linux, the parameters are *OS, browser, protocol, cpu, dbms, and java_version*. Practical testing problems may be more complex than this somewhat contrived example, and may have multiple constraints among parameters. A variety of approaches can be used to deal with this type of problem:

Split test suite: The simplest and perhaps most obvious method is to switch from a single-configuration test suite to one test suite for each combination of parameters that control the applicability of other parameters. In this case, there would be one test suite for Linux and one for Windows systems. This setup is easy to accomplish, but results in some duplicate combinations. For example, the same three-way combinations for *browser, protocol*, and *dbms* will occur in both test suites. The situation is helped a bit if splitting the tests into two separate arrays means two covering arrays for $n - 1$ parameters instead of one covering array for n parameters, and we will have fewer tests with one less parameter to cover. But since the number of tests grows with log n, the number of tests for $n - 1$ parameters is just slightly smaller than for n. In general, therefore, splitting the problem into two test suites will result in almost twice the number of tests. For example, for $t = 3$, $v = 3$, a covering array for 10 parameters has 66 tests, and for nine parameters, there are 62 tests.

Covering arrays with shielding parameters: It is also possible to use an algorithm that allows the specification of "shielding" parameters [40]. In the example above, *dot_net_version* does not apply where the OS parameter is Linux. A parameter that does not always appear (in this case, *dot_net_version*) is called a *dependent* parameter, the parameter that controls the dependent parameter used is called the *shielding* parameter, and the values of the shielding parameter that control the use of the dependent parameter are called controlling values (here, OS = Linux). This method prevents the generation of a large number of duplicate combinations.

However, this approach requires modification of the covering array generation algorithm, and the shielded parameter approach is not yet implemented in most covering array tools.

Combine parameters: An alternative approach is to combine parameters that do not apply to all configurations with other parameters, then to use constraints. This is essentially a way of using the "shielded parameters" concept without requiring a modified covering array algorithm. In this case, "*java_version*" and "*dot_net_version*" could be combined into a single "*platform_version*." Constraints could be used to prevent the occurrence of invalid platform versions. For example, if the Java versions being included in the tests are 1.6 and 1.7, and .Net versions are 3 and 4, then the following parameter can be established:

```
platform_version: {java1.6, java1.7, dot_net3, dot_
net4}
constraint: (OS="Linux" => platform_version="java1.6"||
platform_version="java1.7")
```

This approach prevents the generation of duplicate three-way combinations for *java_version*, *protocol*, and *dbms* in both test suites. That is, a particular three-way combination of these parameters will occur in association with at least one combination, but not necessarily with both OSes in the test suite. The advantage of this approach is that it can be used with any covering array tool that implements constraints. It also produces reasonably compact covering arrays that are suitable for practical testing. The data modeling approaches for handling some of the problems discussed here are discussed in Chapter 5.

3.4 COST AND PRACTICAL CONSIDERATIONS

Applying combinatorial methods for testing configurations can be highly cost-effective. Most software applications are required to run on a variety of systems, and they must work correctly on different combinations of OS, browser, hardware platform, user interface, and other variables. The constraints among parameter values are very common in practical testing. Depending on the constraints needed, the size of the test suite may either decrease or increase with constraints because the covering array algorithm has less opportunity to compress combinations in tests. The increase in test set size is not always significant, but must be kept in mind in initial planning.

One of the key questions in any software assurance effort concerns how many tests are required. Unfortunately, there is no general formula to compute the size of a covering array with constraints and parameters with varying numbers of values (mixed level arrays). If all parameters have the same number of values, or at least have little variation among values (e.g., mostly binary with a few having three values), then tables of covering arrays may be used to determine the number of tests needed in advance. See Appendix C for links to precomputed covering arrays and best-known sizes of arrays for particular configurations. For mixed level arrays, particularly where there is significant variation among the number of values per parameter, the situation is more complex. If v_l is the least number of values for n parameters and if v_m is the greatest, the number of tests will lie somewhere between the size of a covering array for $(v_l)^n$ and $(v_m)^n$, but the interpolation is not linear. For example, a 3-way array for a configuration of $2^8 10^2$ has 375 tests, while the 2^{10} configuration has 66 tests and the 10^{10} configuration has 2367 tests. The situation is even more complex with more variability among the parameter values, or in the presence of constraints; so, generally there is no practical way to determine the number of tests without running the covering array generator tool.

3.5 CHAPTER SUMMARY

Configuration testing is probably the most commonly used application of combinatorial methods in software testing. Whenever an application has roughly five or more configurable attributes, a covering array is likely to make testing more efficient. The configurable attributes usually have a small number of possible values, which is an ideal situation for combinatorial methods. Since the number of t-way tests is proportional to $v^t \log n$, for n parameters with v values, as long as configurable attributes have less than around 10 possible values, the number of tests generated will probably be reasonable. The real-world testing problem introduced in Section 3.2 is a fairly typical size where 4-way interactions can be tested with a few hundred tests.

Since many systems have certain configurations that may not be of interest (such as the Internet Explorer browser on a Linux system), constraints are an important consideration in any type of testing. With combinatorial methods, it is important that the covering array generator allows for the inclusion of constraints so that all relevant interactions are tested, and important information is not lost because a test contains an impossible combination. Constraints may exist between parameter values or may even affect the presence of certain parameters in testing. An example of the former is the constraint "OS = Linux => browser ≠ IE," where the value of the "OS" parameter affects the value of the "browser" parameter. The second type of constraint involves what have been termed "shielding parameters," such as the case where "OS = Linux" means that the parameter "dot_net_version" should not appear in a test, but if "OS = Windows," the test may have both a .Net version and a Java version. A practical workaround for this situation is to merge the dependent parameter into an abstract parameter such as "platform" and then to use constraints among values to prevent the production of tests with nonexistent configurations.

REVIEW

Q: If a system has six configuration parameters with four values each, how many possible configurations are there?

A: $4^6 = 4096$.

Q: If a system has six binary parameters, two parameters with three values each, four parameters with five values each, and four parameters with 10 values each, how many possible configurations are there?

A: $2^6 3^2 5^4 10^4 = 3{,}600{,}000{,}000$.

Q: For the system in the first problem, six parameters with four values each, how many pairs of parameters are there? How many triplets?

A: C(6, 2) = 6!/4!2! = 15 pairs. C(6, 3) = 6!/3!3! = 20 triplets.

Q: For the system in the second problem, how many pairs of parameters are there? How many triplets?

A: C(16, 2) = 120 pairs. C(16, 3) = 560 triplets.

Q: For the system in the first problem, six parameters with four values each, how many 2-way settings of parameters are there? How many 3-way settings?

A: $4^2 \times C(6,2) = 240$ two-way settings. $4^3 \times C(6,3) = 1280$ three-way settings.

CHAPTER 4

Input Testing

As noted in the introduction, the key advantage of combinatorial testing derives from the interaction rule: all, or nearly all, software failures involve interactions of only a few parameters. Using combinatorial testing to select configurations can make testing more efficient, but it can be even more effective when used to select input parameter values. Traditionally testers develop scenarios of how an application will be used, then select inputs that will exercise each of the application features using representative values, normally supplemented with extreme values to test the performance and reliability. The problem with this often *ad hoc* approach is that unusual combinations will usually be missed, so that a system may pass all tests and may work well under normal circumstances, but may eventually encounter a combination of inputs that it fails to process correctly. By testing all *t*-way combinations, for some specified level of *t*, combinatorial testing can help to avoid this type of situation.

4.1 PARTITIONING THE INPUT SPACE

To get a sense of the problem, we will consider a simple example. The system under test is an access control module that implements the following policy:

Access is allowed if and only if

- The subject is an employee
  - AND the current time is between 9 a.m. and 5 p.m.
  - AND it is not a weekend

- OR the subject is an employee with a special authorization code
- OR the subject is an auditor AND the time is between 9 a.m. and 5 p.m. (not constrained to weekdays)

The input parameters for this module are shown in Figure 4.1. In an actual implementation, the values for a particular access attempt would be passed to a module that returns a "grant" or "deny" access decision, using a function call such as "`access _ decision(emp, time, day, auth, aud).`"

Our task is to develop a covering array of tests for these inputs. The first step will be to develop a table of parameters and possible values, similar to that in Section 3.1 in the previous chapter. The only difference is that, in this case, we are dealing with input parameters rather than configuration options. For the most part, the task is simple: we just take the values directly from the specifications or code, as shown in Table 4.1. Several parameters are boolean, and we will use 0 and 1 for false and true values, respectively. For days of the week, there are only seven values; so, all these can be used. However, hour of the day presents a problem. Recall that the number of tests generated for *n* parameters grows proportional to v^t where v is the number of values and t is the interaction level (2-way through 6-way). For all boolean values and 4-way testing, v^t is 2^4. But consider what happens to the test set size with a large number of possible values, such as 24 h, since $24^4 = 331{,}736$. Even worse in this example, time is given in

```
emp:  boolean;
time:  0..1440;   // time in minutes
day:   {m,tu,w,th,f,sa,su};
auth: boolean;
aud:  boolean;
```

FIGURE 4.1 Access control module input parameters.

TABLE 4.1 Parameters and Values for Access Control Example

Parameter	Values
emp	0, 1
time	??
day	`m,tu,w,th,f,sa,su`
auth	0, 1
aud	0, 1

minutes, which would obviously be completely intractable. Therefore, we must select representative values for the hour parameter. This problem occurs in all types of testing, not just with combinatorial methods, and good methods have been developed to deal with it. Most testers are already familiar with one or more of these: *category* [153] *or equivalence* [165] *partitioning* and *boundary value analysis*. These methods are reviewed here to introduce the examples. A much more systematic treatment, in the context of data modeling, is provided in Section 5.6. Additional background on these methods can be found in software testing texts such as Ammann and Offutt [4], Beizer [13], Copeland [57], Mathur [128], and Myers [139].

Both these intuitively appealing methods will produce a smaller set of values that should be adequate for testing purposes, by dividing the possible values into partitions that are meaningful for the program being tested. One value is selected for each partition. The objective is to partition the input space such that *any value selected from the partition will affect the program under test in the same way as any other value in the partition.* That is, from a testing standpoint, the values in a partition are equivalent (hence the name "equivalence class"). Thus, ideally if a test case contains a parameter *x* that has value *y*, replacing *y* with any other value from the partition will not affect the test case result. This ideal may not always be achieved in practice.

How should the partitions be determined? One obvious, but not necessarily good, approach is to simply select values from various points on the range of a variable. For example, if capacity can range from 0 to 20,000, it might seem sensible to select 0, 10,000, and 20,000 as possible values. But this approach is likely to miss important cases that depend on the specific requirements of the system under test. Engineering judgment is involved, but partitions are usually best determined from the specification. In this example, 9 a.m. and 5 p.m. are significant; so, 0540 (9 h past midnight in minutes) and 1020 (17 h past midnight in minutes) could be used to determine the appropriate partitions.

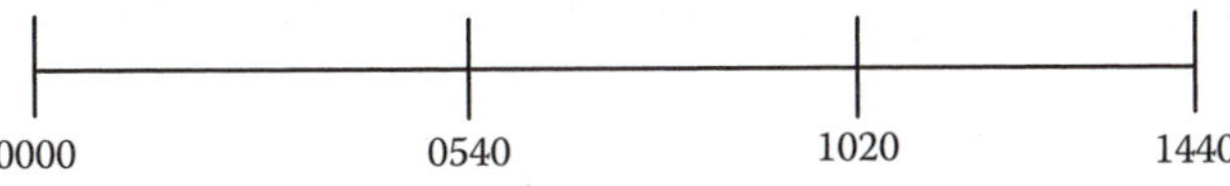

Ideally, the program should behave the same for any of the times within the partitions; it should not matter whether the time is 4:00 a.m.

Use a maximum of 8–10 values per parameter to keep testing tractable.

---

or 7:03 a.m., for example because the specification treats both these times the same. Similarly, it should not matter which time between the hours of 9 a.m. and 5 p.m. is chosen; the access control program should behave the same for 10:20 a.m. and 2:33 p.m. because these times are treated the same in the specification. One common strategy, *boundary value analysis,* is to select test values at each boundary and at the smallest possible unit on either side of the boundary, for three values per boundary. The intuition, backed by empirical research, is that errors are more likely at boundary conditions because errors in programming may be made at these points. For example, if the requirements for automated teller machine software say that a withdrawal should not be allowed to exceed $300, a programming error such as the following could occur:

```
if (amount > 0 && amount < 300) {
        //process withdrawal
} else {
        //error message
}
```

Here, the second condition should have been "`amount <= 300;`" so, a test case that includes the value `amount = 300` can detect the error, but a test with `amount = 305` would not detect the error. Generally, it is also desirable to test the extremes of ranges. One possible selection of values for the time parameter would then be: 0000, 0539, 0540, 0541, 1019, 1020, 1021, and 1440. More values would be better, but the tester may believe that this is the most effective set for the available time budget. With this selection, the total number of combinations is $2 \times 8 \times 7 \times 2 \times 2 = 448$. Generating covering arrays for $t = 2$ through 4 results in the following (Table 4.2) number of tests.

It is important to keep in mind that parameters may not always appear in a single function call, such as our example `access_decision(emp, time, day, auth, aud)`. Sometimes, inputs to a particular operation

TABLE 4.2 Number of Tests for Access Control Example

t	**# Tests**
2	56
3	112
4	224

may be spread through many lines of code in a program. For instance, consider an automated teller machine processing input from a user and the user's ATM card. The code may contain a series of calls such as the following:

```
get_acct_num(); //read acct number from card
get_PIN();      //read PIN from keyboard
get_tran_type(); //read transaction type, withdrawal
  or deposit
get_amt();      //read transaction amount from
  keyboard
process_tran(); //process transaction
```

In this case, a series of values will be established in the memory before finally being processed. So, account number, PIN, transaction type, and amount are all parameters used in tests, but they are being entered one at a time instead of all at once. This situation is common in real-world systems.

4.2 INPUT VARIABLES VERSUS TEST PARAMETERS

For some applications, we test combinations of input characteristics, not just inputs.

In the example above, we assumed that the parameters to be included in the tests were taken from function calls in the program, $f\,(p_1, p_2, \ldots, p_n)$, where each parameter had defined values or a range of values. In many cases, it will not be so obvious how to identify what should be included in the covering array and tests. The classic Ostrand and Balcer [153] category partitioning paper illustrates this common situation with the example of a "find" command, which takes the user input of a string and a file name and locates all lines containing the string. The format of the command is "find <*string*> <*filename* >, where <*string*> is one or more quoted strings of characters such as "john," "john smith," or "john" "smith." Search strings may include the escape character (backslash) for quotes, to select strings with embedded quotes in the file, such as "\"john\"" to report the presence of lines containing *john* in quotes within the file. The command displays any lines containing one or more of the strings. This command has only two input variables, *string* and *filename*; so, is combinatorial testing really useful here?

In fact, combinatorial methods can be highly effective for this common type of testing problem. To check the "find" command, testers will want to ensure that it handles inputs correctly. The input variables in this case are *string* and *filename*, but it is common to refer to such variables as *parameters*. We will distinguish between the two here, but follow conventional practice where the distinction is clear. The *test parameters* or *abstract parameters* identify characteristics of the command input variables. So, in this case, the *test parameters* are different from the two *input parameters*, *string* and *filename*. For example, the *string* input has characteristics such as length and presence of embedded blanks. Clearly, there are many ways to select test parameters; so, engineering judgment must be used to determine what is most important. One selection could be the following, where *file_length* is the length in characters of the file being searched:

String length: {0, 1, 1..*file_length*, >*file_length*}

Quotes: {yes, no, improperly formatted quotes}

Blanks: {0, 1, >1}

Embedded quotes: {0, 1, 1 escaped, 1 not escaped}

Filename: {valid, invalid}

Strings in command line: {0, 1, >1}

String presence in file: {0, 1, >1}

For these seven test parameters, we have $2^1 3^4 4^2 = 2592$ possible combinations of test parameter values. If we choose to test all 2-way interactions, we need only 19 tests. For 3- and 4-way combinations, we need only 67 and 218 tests, respectively. Since the number of tests grows only as log n for n parameters, we can do very thorough testing at relatively low cost for problems such as this. That is, we can include a large number of characteristics to be used as test parameters without significantly increasing the test burden. In the problem above, if we used only the first four of the test parameters, instead of all seven, the number of tests required for $t = 2$, 3, and 4, respectively are 16, 54, and 144. Using all seven characteristics means much more thorough testing with relatively little increase in the test set size.

When testing combinations of input characteristics as above, we must be careful that the test set captures enough important cases. For the find command, testing 3-way or 4-way combinations of the seven characteristics should be an excellent sample of test cases that can detect problems.

That is, the tests will include both valid and invalid strings. In some cases, there may be a need to ensure the presence of test cases with a number of specific characteristics. For example, passwords may be required to (1) exceed a certain length, (2) contain numerics, and (3) contain special characters. A 2-way covering array might not include any valid cases because it contains all pairs but three characteristics must be true to constitute a valid test case. We may need to supplement the covering array with some additional tests in this case. Section 5.4.3 discusses this situation in more detail, along with ways to deal with it. A good case study that illustrates the use of some of these techniques can be found in Borazjany et al. [21].

4.3 FAULT TYPE AND DETECTABILITY

Consider the code snippet introduced in Figure 1.1 of Chapter 1 again. As seen below, if two boolean conditions are true, faulty code is executed, resulting in a failure

```
if (pressure < 10) {
   //do something
   if (volume > 300) {
            //faulty code! BOOM!
      } else {
            //good code, no problem
      }
} else {
  //do something else
}
```

In this case, the branches `pressure < 10` and `volume > 300` are correct and the fault occurs in the code that is reached when these conditions are true. Thus, any covering array with values for pressure and volume that will make the conditions true can detect the problem. But consider another type of fault, in which branching statements may be faulty. The difference between these two types of faults is illustrated below, which we will refer to as (a) *code block faults* and (b) *condition faults*:

Example 1

a. Code block fault example

```
if (correct condition) {faulty code}
else                   {correct code}
```

b. Condition fault example

```
if (faulty condition)    {correct code}
else                     {correct code}
```

Now, suppose the code is as follows:

Example 2

```
if ((a || !b) && c) {faulty code}
else                {correct code}
```

Condition faults are much more difficult to detect than code block faults.

In this case, a 2-way covering array that includes values for a, b, and c is guaranteed to trigger the faulty code, since a branch to the faulty code occurs if either a && c or !b && c is true. A 2-way array will contain both these conditions; so, only pairs of values are needed even though the branch condition contains three variables. Suppose that the fault is not in the code block that follows from the branch but in the branch condition itself, as shown in the following code block. In this case, block 1 should be executed when (a || !b) && c evaluates to true and block 2 should be executed in all other cases, but a programming error has replaced || with &&.

```
if ((a && !b) && c) {block 1, correct code}
else                {block 2, different correct code}
```

A 2-way covering array may fail to detect the error. A quick analysis shows that the two expressions (a && !b) && c and (a || !b) && c evaluate differently for two value settings: a,b,c = 0,0,1 and a,b,c = 1,1,1. A 2-way array is certain to include all pairs of these values, but not necessarily all three values in the same test. A 3-way array would be needed to ensure detecting the error because it would be guaranteed to include a,b,c = 0,0,1 and a,b,c = 1,1,1, either of which will detect the error.

Detecting condition faults can be extremely challenging. Experimental evaluations of the effectiveness of pairwise (2-way) combinatorial testing [9] show the difficulty of detecting condition faults. Using a set of 20 complex boolean expressions that have been used in other testing studies (see

Refs. [9,10] or [214] for a complete list of expressions), detection was evaluated for five different types of seeded faults. For the full set of randomly seeded faults, pairwise testing had an effectiveness of only 28%, although this was partially because different types of faults occurred with different frequency. For the five types of faults, detection effectiveness was only 22% for one type, but the other four types ranged from 46% to 73%, averaging 51% across all the types. This is considerably below the occurrence rates of 2-way interaction failures reported in Section 4.1, which reflect empirical data on failures that result from a combination of condition faults and code block faults. Even 6-way combinations are not likely to detect all errors in complex conditions. A study [202] of fault detection effectiveness for expressions of 5–15 boolean variables found detection rates for randomly generated faults as shown in Figure 4.2 (2000 trials; 200 per set). Note that even for 6-way combinations, fault detection was just above 80%.

How can we reconcile these results with the demonstrated effectiveness of combinatorial testing? First, note that the expressions used in this study were quite complex, involving up to 15 variables. Also consider that the software nearly always includes code blocks interspersed with nested conditionals, often several levels deep. Furthermore, the input variables

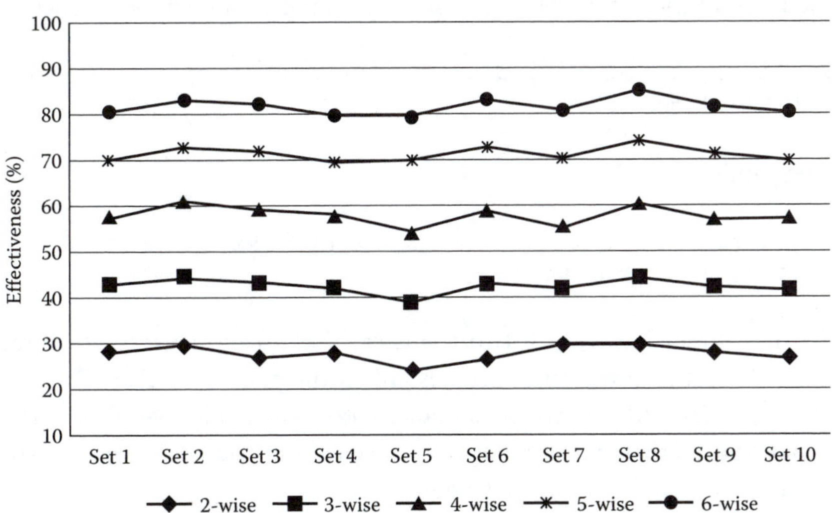

FIGURE 4.2 Effectiveness of *t*-way testing for expressions of 5–15 boolean variables. (Adapted from Vilkomir, S., O. Starov, and R. Bhambroo. Evaluation of *t*-wise approach for testing logical expressions in software, In *Software Testing, Verification and Validation (ICST), 2013 IEEE Sixth International Conference on*, Luxembourg, IEEE, 2013.)

often used in covering arrays are not used directly in conditions internal to the program. Their values may be used in computing other values that are then propagated to the variables in the boolean conditions inside the program, and using high-strength covering arrays of input values in testing may be sufficient for a high rate of error detection. Nevertheless, the results in Ref. [202] are important because they illustrate an additional consideration in using combinatorial methods. For high assurance, it may be necessary to inspect conditionals in the code (if the source code is available) and to determine the correctness of branching conditions through nontesting means, such as formally mapping conditionals to program specifications.

What do these observations mean for practical testing, and what interaction strengths are needed to detect condition faults that occur in actual product software? In general, code with complex conditions may require higher strength (higher level of *t*-way combinations) testing, which is not surprising. But it also helps to explain why relatively low-strength covering arrays may be so effective. Although the condition in Example 1 above includes three terms, it expands to a disjunctive normal form (DNF) of `a && c || b && c`; so, only two terms are needed to branch into the faulty code. Even a more complex expression with many different terms, such as

```
if ((a || b) && c || d && e && (!f || g) || !a && (d
  || h || j))...
```

expands to

```
if (a && c || b && c || d && !a || h && !a || d && e
  && g || d && e && !f)...
```

that has three clauses with two terms each, and two clauses with three terms. Note that a test that includes any of the pairs `[a c], [b c], [d !a], [h !a]` will trigger a branch into code that follows this conditional. Thus, if that code is faulty, a 2-way covering array will cause it to be executed so that the error can be detected.

These observations lead us to an approach for detecting code block faults: given any complex condition, *P*, convert *P* to DNF, then let *t* equal the smallest number of literals in any term. A *t*-way covering array will then include at least one test in which the conditional will evaluate to true if the conditional uses input parameter values, thus branching into the code that follows the conditional. For example, convert `((a || !b)`

`&& c)` to `(a && c) || (!b && c)`; then $t = 2$. Again, however, an important caveat to this approach is that in most software, conditions are nested, interspersed with blocks of the code, so that the relationship between the code block faults and condition faults is complex. A faulty condition may branch into a section of code that is not correct for that condition, which then computes values that may be used in a nested conditional statement, and so on. Unless the conditional is using input parameter values directly, it is not certain that the inputs will propagate to the parameters in the conditional. This approach is a heuristic that is helpful in estimating the interaction strength (level of t) that can be used, but not a guarantee.

4.4 BUILDING TESTS TO MATCH AN OPERATIONAL PROFILE

Combinatorial test sets can approximate an operational profile with some loss of efficiency.

Many test projects require the use of an operational profile [135,136], which attempts to use the same probability distribution of inputs for test data as occurs in live system operation. For example, if a web banking system typically receives 40% balance inquiries, 40% payroll deposit transactions, and 20% bill-pay transactions, then the test data would include these three transaction types in approximately the same proportion. Similarly, an operational profile may be applied to the input data in each transaction, and the test data would be matched to this distribution. For example, an input partition for the "amount" field in the bill-pay transaction might include inputs of 96% with amounts under the user's balance, 3% with insufficient funds, and 1% zero amounts (user error), similar to the proportion of values that the bank experiences in day-to-day use of their system. How can the operational profile approach be used in conjunction with combinatorial testing?

One way we can approximate an operational profile for some problems is to assign sets of values to variables in proportion to their occurrence in the operational profile, if the chances of their occurrence in the input are independent of each other. For example, if we have five binary variables, $a..e$, where a and b have value 0 two-thirds of the time and value 1 a third of the time, and the rest of the binary variables have 0, 1 with equal chance. Then use this as an input to ACTS, assigning 0 and 1 in proportion to the occurrence of 0 for *a* and *b* (2/3), and assigning 2 in proportion to the occurrence of 1 (1/3):

a: 0,1,2

b: 0,1,2

c: 0,1

d: 0,1

e: 0,1

In the covering array, change 1 to 0 for variables *a* and *b*, then change 2 to 1:

a b c d e		***a b c d e***
0,0,0,0,0		0,0,0,0,0
0,1,1,1,1		0,0,1,1,1
0,2,0,1,0		0,1,0,1,0
1,0,1,0,1		0,0,1,0,1
1,1,0,0,0	becomes →	0,0,0,0,0
1,2,1,1,1		0,1,1,1,1
2,0,0,1,1		1,0,0,1,1
2,1,1,0,0		1,0,1,0,0
2,2,*,0,*		1,1,*,0,*

We will have inputs where $a, b = 0, 0$ 4/9 of the time, $a, b = 0, 1$ 2/9 of the time, and so on. It is just an approximation to the correct distribution though, since the distribution is not quite right for some combinations, for example, $b, c = 1, 0$ only 1/9, instead of 1/6, depending on what we do with the * in the last row. This approach would obviously be a lot messier if we were trying to do distributions with lots of values per variable. There are ways to make this more efficient, but we should probably stick with the things we can do using ACTS, and we should not implement new algorithms since practical problems will require constraint handling.

Limitations: Fine-grained control of the distribution of test values is not practical with this approach because it relies on using multiple values that are then mapped into a smaller set of desired values to produce the distribution. Thus, if the desired distribution is 60/20/20 for three values of parameter P1, we can specify the input to the covering array generator as follows:

P1: *a*1, *a*2, *a*3, *b*, *c*.

Then the covering array will have approximately 3 times as many values of "*a*" for P1 if we map *a*1, *a*2, and *a*3 to *a*. We will refer to the values *a*1, *a*2, and *a*3 as "temporary" values, which are mapped to the "actual" value *a*. A distribution such as 45/25/20/10 for four values *a*, *b*, *c*, and *d* would be much more difficult to approximate. It requires that value *a* appears in the covering array 4.5 times as frequently as value *d*, value *b* appears 2.5 times for each occurrence of *d*, and *c* must be twice as common as *d*. Since we are obviously limited to whole numbers of value occurrences, the way to do this would be as follows:

P1: *a*1, *a*2, *a*3, *a*4, *a*5, *a*6, *a*7, *a*8, *a*9, *b*1, *b*2, *b*3, *b*4, *b*5, *c*1, *c*2, *d*.

Unfortunately, this results in 17 temporary values for parameter P1. Recall from Chapter 1 that the number of tests is proportional to v^t; so, even if $t = 2$ or $t = 3$, the resulting covering array of tests will be extremely large. A more practical approach to this problem is to trade some of the precision in the distribution for a smaller test set. If we are willing to accept an approximate distribution of 40/20/20/10 instead of 45/25/20/10, then we reduce the number of values for P1 to 9 instead of 17 (*a*1, *a*2, *a*3, *a*4, *b*1, *b*2, *c*1, *c*2, *d*). One heuristic that helps to make it more practical to generate test arrays meeting an operational distribution is to require that the proportions of different values must all be divisible by at least 10, ensuring that not more than 10 temporary values are used. For example, a 60/20/10/10 distribution can be produced with six for the first value, two for the second value, and so on. Of course, limiting temporary values to 10 or less means that the actual values must be constrained to significantly <10, depending on the distribution being modeled. Once again, engineering judgment is required to find a trade-off that works for the problem at hand.

We also note that operational profile testing is focused on approximating the type and the number of inputs normally encountered, whereas combinatorial testing's forte is exercising the very rare cases that normal testing might miss. An additional complication is that not all failures have the same consequence in terms of economic or other impact. The more commonly used functions of the system may be much more important to a company's revenue, for example, because of the large number of customers impacted when one of them fails. Such considerations argue for the need to consider the operational distribution in test planning, looking at the cost of failure for different functions [214,215]. For example, a retail operation may place a higher priority on customer purchase transactions than on item return, on the basis of both volume and impact on revenue. In this case, it makes sense to do more testing of purchase transactions, reflecting the operational distribution of transaction types. Combinatorial testing would then be applied to testing of purchase transactions to detect obscure input combinations that might cause a failure. Very heavily used transaction types are eventually likely to encounter almost any combination; so, it is important to find these rare cases in testing.

4.5 SCALING CONSIDERATIONS

The larger the system, the greater the benefit from combinatorial testing.

With some of the examples discussed previously, the advantage over exhaustive testing is not large, because of the small number of parameters. What happens with really big problems? For larger problems, the advantages of combinatorial testing can be spectacular. For example, consider the problem of testing the software that processes switch settings for the panel [140] shown in Figure 4.3. There are 34 switches, which can be either on or off, for a total of $2^{34} = 1.7 \times 10^{10}$ possible settings. Clearly, we cannot test 17 billion possible settings, but all 3-way interactions can be tested with only 33 tests, and all 4-way interactions can be tested with only 85 tests. This may seem surprising at first, but it results from the fact that every test of 34 parameters contains C(34, 3) = 5984 3-way and C(34, 4) = 46,376 4-way and combinations.

This example illustrates the fact that the testing efficiency gain from combinatorial methods is much greater with larger problems. Recall from Section 4.2 that the number of tests required for n parameters with v values increases as $v^t \log n$ for t-way testing, but exhaustive testing for the same problem would require v^n tests. Table 4.3 shows the sizes of 2-way and 4-way covering arrays for different levels of v with 10 through 50 variables.

FIGURE 4.3 Panel with 34 switches. (Courtesy of National Aeronautics and Space Administration, Meteorological Measurement System, http://geo.arc.nasa.gov/sgg/mms/Integration/dc8/mission_manager.htm.)

TABLE 4.3 2- and 4-Way Covering Array Sizes Compared with Exhaustive Tests for Various Values of n and v

	$v = 2$			$v = 4$			$v = 6$		
n	2-Way Covering Array	4-Way Covering Array	Exhaustive	2-Way Covering Array	4-Way Covering Array	Exhaustive	2-Way Covering Array	4-Way Covering Array	Exhaustive
10	8	41	1024	29	725	10,48,576	63	3713	6.046e+7
20	10	65	1,048,576	37	1165	1.099e+12	79	6015	3.656e+15
30	11	80	1.073e+9	41	1448	1.1529e+18	86	7473	2.210e+23
40	11	90	1.099e+12	44	1661	1.2089e+24	94	8550	1.336e+31
50	11	98	1.125e+15	46	1839	1.267e+30	99	9466	8.082e+38

Note the logarithmic growth of covering array sizes with increasing values of n, and the fact that the covering arrays are tiny compared with what would be required for exhaustive testing.

4.6 COST AND PRACTICAL CONSIDERATIONS

Combinatorial methods can be highly effective and reduce the cost of testing substantially. For example, Justin Hunter has applied these methods to a wide variety of test problems and consistently found both lower cost and more rapid error detection [101]. Although this book focuses on software testing, the tools and methods described here have also been found useful for problems such as analyzing engineering problems [81] and modeling and simulation [106]. Many problems where the interaction between components or variables is important can benefit from combinatorial methods, and readers are likely to identify some of their own after gaining familiarity with the techniques introduced in this book.

As with most aspects of engineering, trade-offs must be considered. Among the most important aspect is the question of when to stop testing, balancing the cost of testing against the risk of failing to discover additional failures. An extensive body of research has been devoted to this topic, and sophisticated models are available for determining when the cost of further testing will exceed the expected benefits [20,122]. The existing models for when to stop testing can also be applied to the combinatorial test approach but there is an additional consideration: What is the appropriate interaction strength to use in this type of testing?

To address these questions, consider the number of tests at different interaction strengths for an avionics software example [103] shown in Figure 4.4. While the number of tests will be different (possibly much smaller than in Figure 4.4) depending on the system under test, the

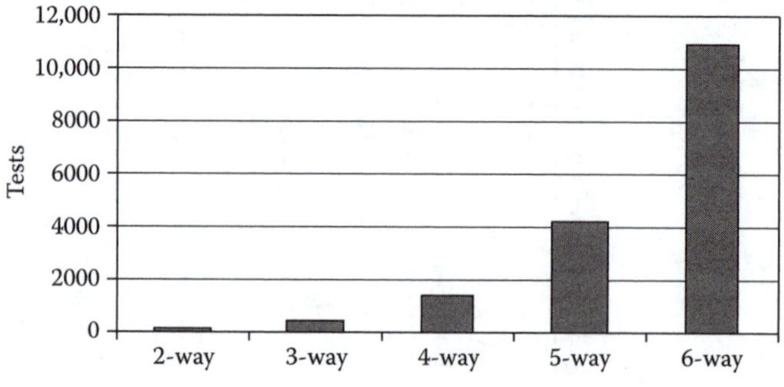

FIGURE 4.4 Number of tests for the avionics example.

magnitude of the difference between the levels of t will be similar to Figure 4.4, because the number of tests grows with v^t for parameters with v values. That is, the number of tests grows with the exponent t; so, we want to use the smallest interaction strength that is appropriate for the problem.

Intuitively, it seems that if no failures are detected by t-way tests, then it may be reasonable to conduct additional testing only for $t + 1$ interactions, but no greater if no additional failures are found at $t + 1$. In the empirical studies of software failures, the number of failures detected at $t > 2$ decreased monotonically with t; so, this heuristic seems to make sense: *start testing using 2-way (pairwise) combinations, continue increasing the interaction strength t until no errors are detected by the* t-way *tests, then (optionally) try* t + 1, *and ensure that no additional errors are detected.* As with other aspects of software development, this guideline is also dependent on resources, time constraints, and cost–benefit considerations.

When applying combinatorial methods to input parameters, the key cost factors are the number of values per parameter, the interaction strength, and the number of parameters. As shown above, the number of tests increases rapidly as the value of t is increased, but the rate of increase depends on the number of values per parameter. Binary variables, with only two values each, result in far fewer tests than parameters with many values each. As a practical matter, when partitioning the input space, it is best to keep the number of values per parameter below 8 or 10 if possible since the number of tests increases with v^t (e.g., consider the difference between $4^3 = 64$ and $11^3 = 1331$).

Since the number of tests increases only logarithmically with the number of parameters, the test set size for a large problem may be only somewhat larger than for a much smaller problem. For example, if a project uses combinatorial testing for a system that has 30 parameters and generates several hundred tests, a larger system with 40–50 parameters may only require a few dozen more tests. Combinatorial methods may generate the best cost–benefit ratio for large systems.

4.7 CHAPTER SUMMARY

1. The key advantage of combinatorial testing derives from the fact that all, or nearly all, software failures appear to involve the interactions of only a few parameters. Generating a covering array of input parameter values allows us to test these interactions, up to a level of 5-way or 6-way combinations, depending on the resources.

2. Practical testing often requires abstracting the possible values of a variable into a small set of equivalence classes. For example, if a variable is a 32-bit integer, it is clearly not possible to test the full range of values in $\pm 2^{31}$. This problem is not unique to combinatorial testing, but occurs in most test methodologies. Simple heuristics and engineering judgment are required to determine the appropriate partitioning of values into equivalence classes, but once this is accomplished, it is possible to generate covering arrays of a few hundred to a few thousand tests for many applications. The thoroughness of coverage will depend on resources and criticality of the application.

REVIEW

Q: When and why is it necessary to partition the input space into discrete values of parameters?

A: When it is not possible to include all possible values of a parameter in tests. This may occur with continuous variables, which may have millions of values. Input space partitioning will also be needed when discrete variables have too many possible values to generate a practical-sized covering array. Generally, the number of values per variable should be kept small, preferably 10 or fewer; so, even a discrete variable with 20 values could generate an excessive number of tests.

Q: When partitioning an input space, what is the goal in selecting boundaries for partitions?

A: Partition the input space such that any value selected from the partition will affect the program under test in the same way as any other value in the partition. In other words, if a test case contains a parameter that has value *y*, replacing *y* with any other value from the partition will not affect the test case result.

Q: What is boundary value analysis?

A: A strategy of selecting partition values on both sides of the boundaries that are significant in the program requirements. For example, if requirements state that accounts with balances of $10,000 and above are processed differently than accounts below this value, then the boundary for the balance variable is 10,000. Test values of 9999, 10,000, and 10,001 would be suitable and likely to reveal errors.

Q: Faults can occur in (a) blocks of sequential code or in (b) branching conditions. Assuming a test oracle exists, which of these two types of faults is likely to be harder to detect and why?

A: Branching conditions (b) may require more tests because we need to show that the correct branch is taken for all possible values of the variables in the branching condition. To detect problems in a block of sequential code, we only need to instantiate to 'true' one of (normally) many conditions that cause a branch into the block (when there is a test oracle to compare results).

Q: How does the number of tests increases with the number of parameters and with the number of values per parameter?

A: The number of tests grows logarithmically with the number of parameters and exponentially (in t) with the number of values per parameter.

Q: What are the practical implications for the way in which the test set size grows with the number of parameters and the number of values per parameter?

A: Since the number of tests increases with the exponent t for v values per parameter, it is better to have fewer values per parameter, and to test t-way interactions for as small a value of t as necessary for the problem. But logarithmic growth with the number of parameters means that large test problems with many parameters are still practical.

CHAPTER 5

Test Parameter Analysis

Eduardo Miranda

THIS CHAPTER ADDRESSES A number of issues encountered in combinatorial testing, namely, what should be included as a test parameter, the selection of parameter values, missing combinations, infeasible combinations, and ineffectual combinations. Different tools address these problems to various degrees using a variety of approaches. All of them however require the identification of the test parameters and the test values involved, so this, and not the tools, will be our focus. The answers given to these problems have arisen from the experience of the author in teaching the method to graduate students in software engineering, from observing its application in the same context, and from a review of the literature. They are not intended to be definitive and they continue to evolve with more extensive use of combinatorial testing. This chapter builds on the concepts introduced in Chapter 3 and those of the classification tree method (CTM) [79] to identify and document the SUT test parameters and their values.

> Analyzing test parameters is the most critical step in the combinatorial testing process.

Section 3.2 briefly distinguished between input variables, test parameters, and parameter values. The first term was used to refer to the SUT's explicit inputs, perhaps as defined by a specification or an API, and the second to any factor, including the explicit inputs, configuration options, and state information that could affect the SUT execution. Test parameter values could be either actual or abstract values such as "1," "John," or "Triangle" or a property that an actual test value must conform to, for example, "A string of length greater than 6" or "An ordered list with no duplicates." We will refer to this collectively as the SUT's input domain.

The CTM is a structured technique for analyzing and documenting the SUT test parameters and values. It was proposed by Matthias Grochtmann and Klaus Grimm in 1993 [79] while working at Daimler-Benz, and extends the ideas pioneered by Ostrand and Balcer in their category partition method (CPM) in 1988 [153]. The method comprises three categories:

- A decomposition criterion for the stepwise identification and refinement of the SUT input domain
- A graphical notation to facilitate communication among people
- A well-formed grammar that supports automatic checking and test case generation

The rest of the chapter is organized as follows: Section 5.1 provides guidelines for the selection of test parameters. Section 5.2 defines missing, infeasible, and ineffectual combinations. Section 5.3 provides a short introduction to the CTM. Section 5.4 elaborates on the modeling method via examples. Sections 5.5 addresses SUT selection and Section 5.6 analyzes different test strategies.

5.1 WHAT SHOULD BE INCLUDED AS A TEST PARAMETER

A test parameter is any factor the tester believes has a bearing on the outcome of a test run. A nonexclusive list includes the SUT input variables, any characteristic of the input values other than its nominal meaning that might force a specified or unspecified execution path, the SUT configuration, the state of the environment in which it executes, and the state of the SUT itself. There is no single criterion for choosing test parameters; it is not a mechanical task. Specification documents are a definitive source since testing has to demonstrate that the software does what it is supposed to do. But they are not very good at specifying how the software should not behave, which is often taken for granted. Other sources of information and inspiration include the SUT signature, mostly in cases like unit testing where specifications at the class or method level are seldom produced; test catalogs [126,154], which list typical test parameters according to the variables' data type; and last but not least, the intuition and knowledge about the SUT implementation the test designer might have.

The SUT input variables: Since the reason an SUT exists is to process its inputs, these should be considered test parameters by definition.

The morphology of the inputs: Sometimes, the attributes of a data type, for example, the cardinality of an array, whether its elements are ordered or not, the length of a password, and the position of a particular character within a sequence can affect the behavior of the SUT. This could just be the unintended by-product of the way an algorithm was coded or the result of constraints imposed on the inputs that need to be verified. Typical examples of the first are failing to check for null strings, whether a character string contains spaces or not, single versus multiple occurrences of a value in an array, and whether values in a list are ordered or not. Test catalogs provide a list of test parameters according to the input type of each variable (see Table 5.1).

TABLE 5.1 Test Catalog

Data Type	**Test Ideas**
Hashing table	A key
	A key that causes a collision
Enumeration	Each enumerated value
	Some value outside the enumerated set
	Some value of a different type
Sequence	Empty
	A single element
	More than one element
	Maximum length (if bounded) or very long
	Longer than maximum length (if bounded)
	Incorrectly terminated
	P occurs at beginning of sequence
	P occurs in interior of sequence
	P occurs at end of sequence
	PP occurs contiguously
	P does not occur in sequence
	pP, where p is a proper prefix of P
	Proper prefix p occurs at end of sequence
	P occurs at beginning of sequence
	P occurs in interior of sequence

Source: Pezze, M. and M. Young. *Software Testing and Analysis—Process, Principles and Techniques*. 2008. Copyright Wiley-VCH Verlag GmbH & Co. KGaA, Hoboken, New Jersey. Reproduced with permission.

Note: P is a sequence element that triggers some processing action when present or absent.

The state of the environment in which the SUT executes: The result of a query is certainly dependent on the data stored in the queried repository. For example, if there are two records that satisfy a query, the SUT might provide a correct answer for the first but fail to find the second. What happens if a repository contains zero, one or two or more occurrences of the same value? In the case of a speech recognition system, the background noise might not be an explicit input to the recognition algorithm, but its level will certainly affect the ability of the system to perform an accurate transcription of the words spoken. In consequence, noise level is a very important test parameter. Other examples could be the amount of memory or other resources available to the SUT, particular software or hardware configurations on which the SUT is executing, and localization data.

The state of the SUT: The state of the SUT refers to the internal information an SUT needs to maintain between successive executions to perform as designed. An SUT that does not need information from a previous execution to respond correctly to a second is called stateless. That is, after the SUT is executed, all local variables and objects that were created or assigned a value as part of the execution are discarded. A *stateful* SUT, instead, retains information from execution to execution and in consequence when presented with the same "inputs" it might produce a different result depending on if it is the first, the second, or any subsequent invocation. From a testing perspective, the usual treatments for this are (1) to "reset" the SUT between tests, this is, for example, the JUnit approach to guarantee repeatable results, (2) to force SUT into a desired state, and once there, to execute the test from that condition, and (3) to define a test sequence, which includes successive executions. If needed, state information shall be incorporated into the combination table by means of abstract values characterizing each state rather than through the inclusion of all intervening variables. This will make the test descriptions resilient to implementation changes while minimizing the size of the test suite.

5.2 COMBINATION ANOMALIES

There are three recurrent issues in the practice of combinatorial testing: missing combinations, infeasible combinations, and ineffectual combinations. We will refer to them as combination anomalies. This section provides a brief review of the subject prior to considering its impact on the classification tree approach. Aspects of these issues have been introduced in previous chapters (e.g., constraints in Section 3.3), but here we discuss them in the context of data modeling and provide more depth. Illustrations and solutions to the problems will be discussed in the context of the examples presented in Section 5.4.

A missing combination is a specific combination required by the test designer for which there is no guarantee that the covering array algorithm will produce, such as a test derived from a particular use case or scenario. Missing combinations occur when the number of intervening test parameters (cardinality of the required combination) is higher than the strength of test suite being generated, for example, all-pairs, all-triples, and so on. For example, if the sought combination includes three test parameters and we are generating a test suite for all pairs, there is no assurance that the required combination will be included. This can be solved by manually including (seeding) the required combinations in the table [50] or by using variable strength relations, which requires higher coverage among a subset of critical parameters (see Appendix D for instructions on how to do this with ACTS).

Infeasible combinations are combinations that are generated by the combinatorial algorithm but are not physically or logically realizable in actual test cases. This situation occurs because the algorithms generating the combinations have no knowledge of the domain semantics. All they have to do is to manipulate symbols with no assigned meaning. The problem with infeasible combinations is that after the test cases containing them are generated, they cannot be simply removed since together with the specific combination of parameter values that renders the combination infeasible, we would be also removing important combinations of values that might not be present in any other test case and in consequence the removal will result in an effective coverage lower than the nominal strength of the generated test suite. There are four fundamental approaches to deal with infeasible combinations, which can be used in isolation or in combination [77]: the abstract parameter method, submodels method, avoid method and the replace method. Whatever technique is chosen to deal with infeasible combinations, a fundamental principle is that the strength of the test suite must be preserved. That is, if a given test suite satisfied the all-pairs coverage with infeasible combinations present, it must satisfy the all-pairs coverage criterion after they are removed. This might be accomplished at the expense of some degree of redundancy in the test suite. Infeasible combinations are usually described to the generating tool via logical expressions (see Section 3.3).

Ineffectual combinations have the same effect as infeasible combinations: loss of effective coverage. There are basically two scenarios: a value forces the brushing aside of other valuable combinations of parameter values included in the same test case and these are never tried. This is the typical scenario when valid and invalid values are mixed in a single test case and processing is terminated upon the detection of the invalid value. This is called input masking [63]. In other cases, certain test parameters are relevant only when a third parameter takes a given value but since all parameters are combined together irrespective of values, some of the combinations are wasted. These parameters are called dependent parameters [40]. An example of this would be a switch selecting an operating mode or an option in a command requiring additional information only applicable to it. The difference between the two cases is that in the first we need to provide all the values, even if they are not going to be used; otherwise, we would not be able to execute the test, but in the second this is not necessary. We would call the value causing the masking or the one that other parameters depend upon a dominant value. Ineffectual combinations might be dealt with by building separate test suites, for example, one test suite for valid values and another for invalid ones for the first case, or by creating two or more combination tables, one for the mandatory parameters and the others for the dependent parameters in the second. The construction can be either automatic or manual through the repeated use of "conventional" combinatorial tools.

5.3 CLASSIFICATION TREE METHOD

The classification tree approach works well with combinatorial methods.

Figure 5.1 provides a pictorial description of the CTM. The basic idea is to partition the input domain of the test object separately under different aspects and then to recombine the different partitions to form test cases [79].

In the CTM, the SUT input domain is partitioned into a hierarchy of test parameters and values (see Figure 5.2). This hierarchical approach has two main advantages: (1) it helps manage the complexity of defining and documenting large input domains and (2) it allows the implementation of hybrid test strategies based on a common description. One could, for example, apply boundary value analysis (BVA) to the lower-level test parameters in the hierarchy, and once their equivalence has been established, use representative values in creating *t*-way combinations with higher up parameters.

The first step in the CTM is to identify all the SUT explicit and implicit factors that might have an impact on its operation. These includes the SUT

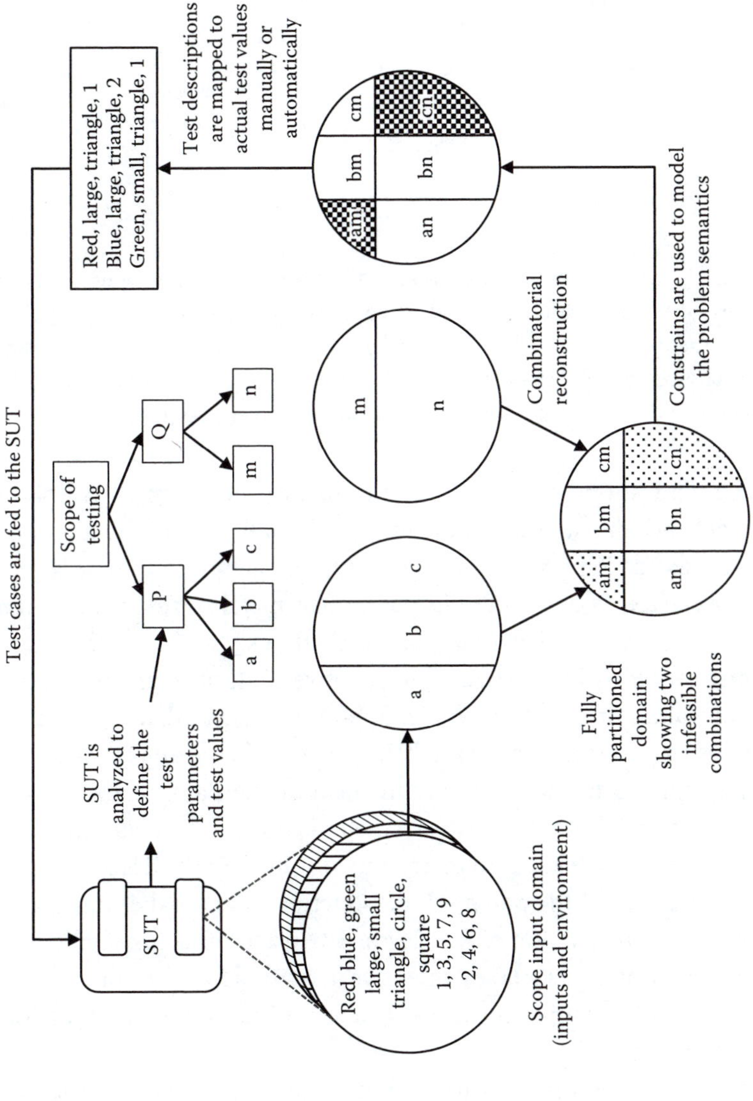

Test cases are fed to the SUT
SUT
SUT is analyzed to define the test parameters and test values
Scope of testing
P
Q
a
b
c
m
n
Red, blue, green
large, small
triangle, circle, square
1, 3, 5, 7, 9
2, 4, 6, 8
Scope input domain (inputs and environment)
a
b
c
m
n
Combinatorial reconstruction
am
bm
cm
an
bn
cn
Fully partitioned domain showing two infeasible combinations
Constrains are used to model the problem semantics
am
bm
cm
an
bn
cn
Test descriptions are mapped to actual test values manually or automatically
Red, large, triangle, 1
Blue, large, triangle, 2
Green, small, triangle, 1

FIGURE 5.1 The classification tree method.

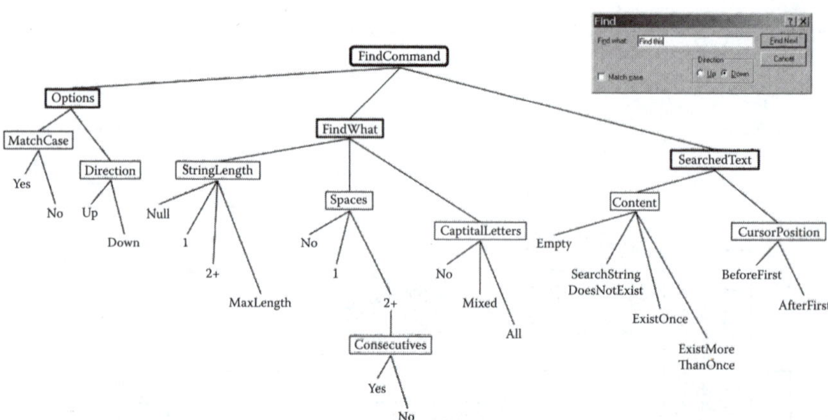

FIGURE 5.2 Classification tree for the "Find" command on the upper right corner. Nodes with thick borders are called compositions and model aggregations of test parameters. Nodes with a thin border are called classifications and model exclusive choices for a test parameter. Nodes without are a border are called classes and correspond to the choices available for a given classification.

inputs, configurations, state of the environment, and internal state. Each of these will become a test parameter. After this first step, the identified parameters are decomposed into a collection of subtest parameters dealing with some relevant aspect of its parent parameter or in a collection of parameter values. In either case, the descendants, values or subparameters must be jointly exhaustive, that is, between all of them, all relevant aspects or values are covered, and mutually exclusive, that is, there are no overlaps among the descendants. The process of decomposition is now recursively applied to the further refinement of test parameters and values. The partitioning of each test parameter is done independently of the partitioning of other test parameters, thus simplifying the decomposition task at the expense of the potential introduction of anomalies during the combinatorial construction of test cases. The trade-off is justified on the basis that partitioning is a labor-intensive activity while the combination and treatment of missing, infeasible, and some ineffectual combinations can be automated.

Classification trees are built using four node types: the root, classifications, compositions, and classes, each of them with a specific meaning and connection rules. The roles each of these node types play in the description of the problem and their connection rules are explained in Table 5.2. The definitions of the nodes and their connections have an unambiguous

TABLE 5.2 Classification Tree Semantic

Modeling Entities	Uses	Example	Allowed Parents	Allowed Descendants
Root	• Denotes the scope of the testing • Consist of relationship	• Engine test (This is what we are testing. We are not testing safety or confort.)		• Compositions • Classifications

Compositions	• To model a test parameter that integrates several other relevant factors
	• To refine a value consisting of several parts
	• Consist of relationship
		• An automobile consists of
		  • Wheels
		  • Engine
		  • Body
		  • Seats
			• Root
			• Composition
			• Class
				• Composition
				• Classification

Classifications	• To model a test parameter that is partitioned into a collectively exhaustive and mutually exclusive sets of values (equivalence classes)
	• Is a relationship
		• An automobile is a
		  • Car
		  • Truck
		  • SUV
		  • Van
			• Root
			• Composition
			• Class
				• Class

Classes	• To model equivalence classes (set of values that elicits the same behavior from the TO) and exemplar values	• *Compact*: Ford Focus, WV Golf • *Midsize*: Chevrolet Malibu, Hyundai Sonata • *Full-size*: Ford Crown, Chevrolet Impala	• Classification	• Composition • Classification

semantics, which permits algebraic manipulation for the purpose of automatically generating test cases and treating of combination anomalies.

The results of the modeling are not unique. As the creators of the method explain, CTM provides a structured and systematized approach to test case design making tests understandable and documentable, but the results will be highly dependent on the tester's experience, knowledge, and creativity. We will discuss the advantages and disadvantages of alternative partitions as we go through the examples in Section 5.4.

The tree in Figure 5.2 describes the input domain for the "Find" command in its upper right corner. The tree nodes correspond to test parameters and values at different levels of abstraction. The thickness of the borders, or the lack of them, surrounding the name of the test parameters differentiate between compositions, classifications, and classes. *Compositions* have a thick border, *classifications* have a thinner one, and *classes* have no border. The root of the tree defines the scope of the test design. By stating the scope as "FindCommand," we have purposefully excluded from the scope other commands embedded in the form such as "?" (help), "Cancel," and "X" (close). The reason for this choice will be discussed in Section 5.5. The second-level nodes highlight the three top-level test parameters: the command options, the search string, and the environment in which the command is to execute. We model the command "Options" as a composition since the choice of "MatchCase" does not exclude the choice of "Direction." The "FindWhat" test parameter, also a composition, models the searched value. Notice that instead of enumerating actual test values such as "FindThis" or "find that," the test parameter was decomposed into three lower-level test parameters corresponding to characteristics of the search string the test designer judged important to test for "StringLength," "Spaces," and "CapitalLetters." These aspects might or might not have been described in the specification document, but in the experience of the test designer, they are things worth including in the test since he has seen developers using spaces as delimiters and programs crashing when presented with a null string. "Null," "1," "2+," and "MaxLength" are values of the "StringLength" test parameter, which is modeled as a classification, since the choice of a particular length precludes the others from being chosen. Classes might model specific values like in the case of the "Null" string or they might denote a set of values deemed equivalent like in the case of the "2+" class, which comprises all strings of length 2 to "MaxLength" – 1. The test designer might choose, for example, to refine the "2+" class using BVA to verify that indeed the processing is equivalent in the range [2, MaxLength – 1]. These will be further discussed in Section 5.6. The third test parameter identified, "SearchedText," is not an explicit input to the command. It is, however, a very important part of the test design since the outcomes of executing the command will depend on the "Content" of the text searched and on the position of the text "CursorPosition" at the time the command is executed.

In general, test parameter values are given in abstract form. For example, to be executable, the parameter value "ExistMoreThanOnce" would need to be transformed in an actual sequence of strings of which at least two must be identical to the one used as value for the "FindWhat" test parameter.

5.4 MODELING METHODOLOGY

In this section, we will illustrate the use of CTM by way of a series of examples. The first is a flexible manufacturing system, which is an extension of the machine vision system used by Grochtmann [78]. Its purpose is to explain the main modeling concepts and how to deal with valid and invalid values to prevent the introduction of ineffectual combinations in a test suite. The second example corresponds to a sound amplifier and is used to further illustrate the handling of ineffectual combinations. The third is a password diagnoser, which addresses the problems of missing and infeasible combinations.

5.4.1 Flexible Manufacturing System Example

The purpose of the flexible manufacturing system is to pick up metal sheets of different shapes, colors, and sizes transported by a conveyor and put them into appropriate bins. The sheets are transported by a conveyor that positions them in front of a camera that feeds the image to a machine vision system. The system classifies them and issues a command to the robotic arm subsystem to pick up the sheet (see Figure 5.3), and move it to the assigned bin. The system is designed to recognize three types of shapes: triangles, circles, and squares; three colors: green, blue, and red; and two sizes: small and large in any combination.

The first step in the process is to select the scope of testing. Are we going to test the whole system, the machine vision subsystem, or the robotic arm subsystem? If we choose the vision subsystem, our concern will be that the system is able to correctly classify the sheets and issue commands. If we choose the robotic arm portion, the starting point will be the reception of a well-formed command, the ability of the system to interpret and execute

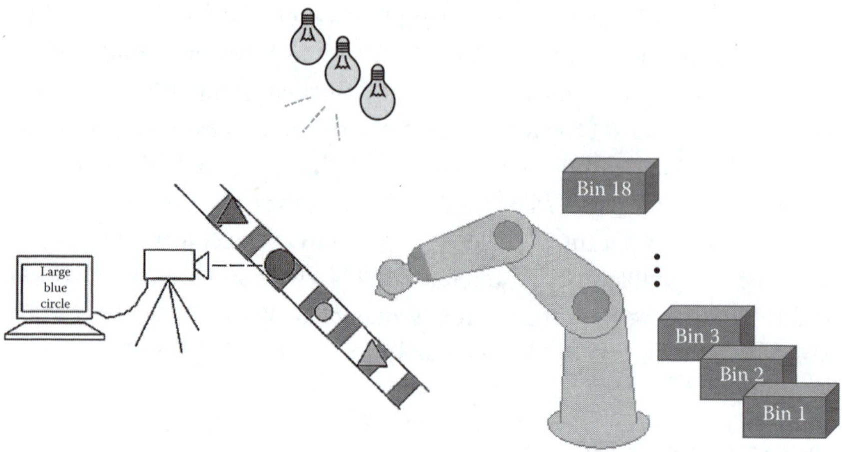

FIGURE 5.3 **(See color insert.)** The flexible manufacturing system.

the command. If we choose the whole system, it will be a combination of the two previous scenarios plus some other cases such as powering-up, shutting down, emergency stop, and so on. For the sake of this example, we have selected the machine vision subsystem component as the SUT. Once this has been decided, the test designer might start by considering the figures to be recognized, an obvious choice, since the purpose of the subsystem is to classify them according to their shape, color, and size as a test parameter. He might also consider environmental conditions such as the room illumination, the conveyor background, and the speed at which the figures move in front of the camera to be relevant test parameters either because a specification called for it or because the test designer suspects the performance of the subsystem might be affected by them. The first cut at the problem is documented by the classification tree in Figure 5.4.

Because each figure is defined by three nonexclusive attributes, shape, color, and size, we model "Figure" as a composition node (see Figure 5.4). Each of the attributes is necessary to define the figure. Contrast this with the room "Lighting," which is modeled as a classification node. If the "Lighting" is "Dim," whatever the definition of "Dim" is, it cannot be "Bright." Classifications define exclusive choices. The "Conveyor" aspect is also modeled as a composition since its effect on the correctness of the classification depends on two coexisting attributes: the conveyor's "Speed" and its "Background."

From a purely semantic perspective, we did not need to introduce the nodes "Figure" and "Conveyor" in the tree (see Figure 5.5), since the root

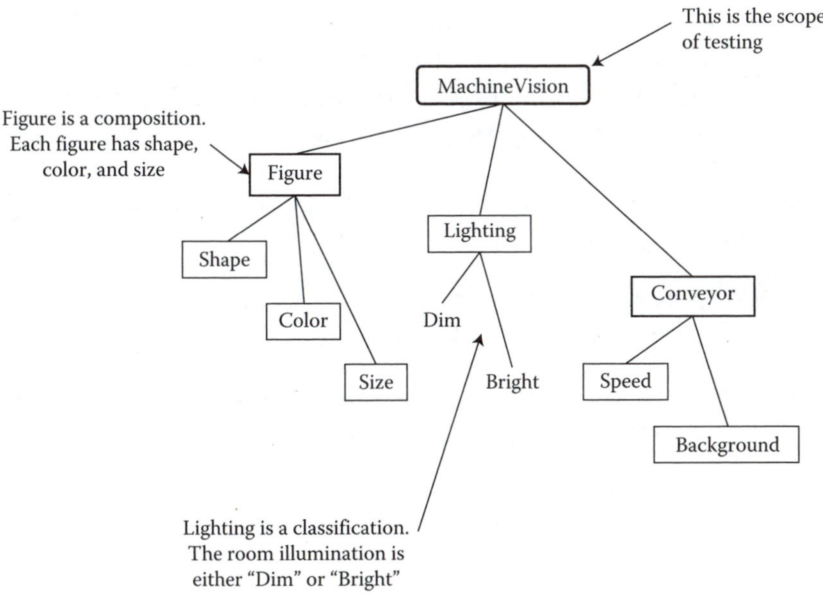

FIGURE 5.4 First cut decomposition for the machine vision subsystem.

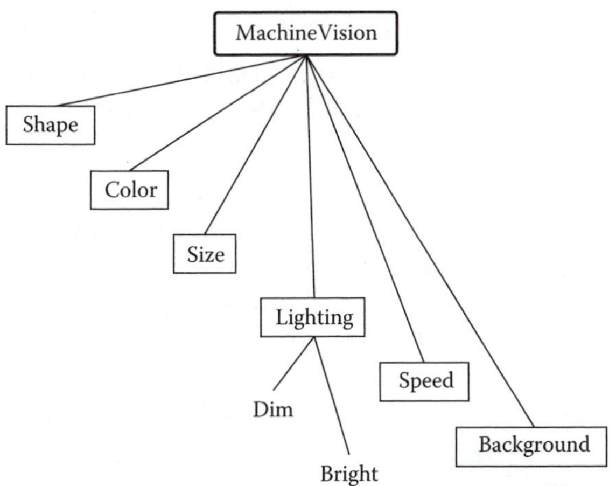

FIGURE 5.5 A semantically equivalent model.

of the tree is also a composition. The representation in Figure 5.4, however, is more descriptive and better communicates the nature of the problem and the test parameters being considered.

Once the first-tier test parameters have been identified, we need to decide whether to assign terminal values to them or to further decompose the parameters into sub-test parameters. Notice that given the hierarchical approach taken, subtest parameters might be employed as parameters in their own right or as values of the parent parameter. In general, a test parameter or a value will be decomposed when

- The entity represented by it is made up of subentities that might affect the output of a test.
- The entity represented by it can be categorized along several dimensions the test designer would like to explore.
- The value poses structural features that are known to be error prone from a programming point of view.

Developing the classification tree is an iterative process.

Figure 5.6 shows that "Shape" might take the values "Triangle," "Circle," and "Square." Shape is modeled as a classification since if a sheet has a "Triangle" form, it cannot have a "Circle" or "Square" shape. Likewise "Color," "Size," "Speed," and "Background" are modeled as classifications. At this point, the test designer might ask himself about different types of triangles. Is the machine vision system capable of recognizing "Equilateral," "Scalene," and "Isosceles" triangles? This might be specified or not, but it is a legitimate question. If the specification requires the recognition of "Equilateral" triangles, then the other two types must not be classified as triangles, so we need to test if this is indeed the case. If the specification assumed all kinds of triangles, then the tester must

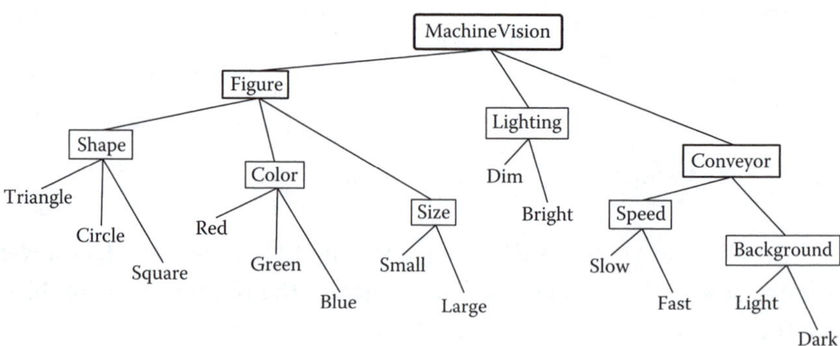

FIGURE 5.6 Second-order refinement of the machine vision subsystem tree.

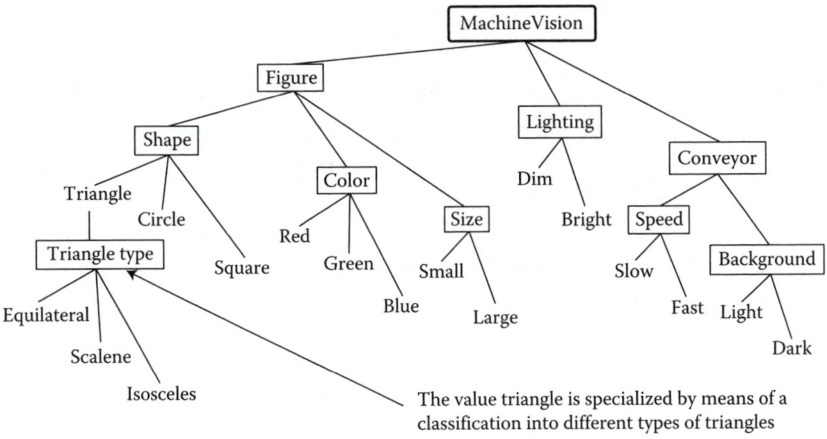

FIGURE 5.7 Tree showing the refinement of a value (class) into more specialized values.

assure that the system does not only work with equilateral, but also with the other types of triangles as well. In consequence, the kind of triangle is a test-relevant aspect that applies to the value "Triangle," which now becomes a sub-test parameter. This is modeled as shown by Figure 5.7.

Before continuing with the refinement of the model, let us illustrate the relationship between the model constructed so far and combinatorial testing. The obvious choice would be to consider "Shape," "Color," "Size," "Lighting," "Speed," and "Background" as a combination table's test parameters and their respective leaf classes as their values. Assuming that the desired strength is to cover all three-way interactions, this will result in the test suite shown by Table 5.3. In previous discussion, we have said that the choice of parameters and values will depend on the test strategy followed. In the example shown in Table 5.3, we chose to treat "Equilateral," "Scalene," and "Isosceles" as a kind of shape and combined them with all the other parameters. Another strategy (see Table 5.4) resulting in fewer test cases would have been to first establish the system was capable of recognizing the three types of triangles under fixed conditions or using an all single values approach, and then to test for the interactions utilizing "Triangle" as an abstract value instead of "Equilateral," "Scalene," and "Isosceles." This is a very important decision, for example, when defining values using BVA since it ought to be remembered that, for a given test strength, the number of test cases generated grows with the product of the number of values. We will further discuss different testing strategies in Section 5.6.

TABLE 5.3 All Triples for the Tree in Figure 5.7

Test Case Number	Shape	Color	Size	Lighting	Speed	Background
1	Equilateral	Red	Small	Dim	Fast	Light
2	Equilateral	Red	Large	Bright	Slow	Dark
3	Equilateral	Green	Small	Bright	Fast	Dark
4	Equilateral	Green	Large	Dim	Slow	Light
5	Equilateral	Blue	Small	Dim	Slow	Dark
6	Equilateral	Blue	Large	Bright	Fast	Light
7	Scalene	Red	Small	Bright	Slow	Light
8	Scalene	Red	Large	Dim	Fast	Dark
9	Scalene	Green	Small	Dim	Fast	Light
10	Scalene	Green	Large	Bright	Slow	Dark
11	Scalene	Blue	Small	Bright	Fast	Dark
12	Scalene	Blue	Large	Dim	Slow	Light
13	Isosceles	Red	Small	Dim	Slow	Dark
14	Isosceles	Red	Large	Bright	Fast	Light
15	Isosceles	Green	Small	Bright	Slow	Light
16	Isosceles	Green	Large	Dim	Fast	Dark
17	Isosceles	Blue	Small	Dim	Fast	Light
18	Isosceles	Blue	Large	Bright	Slow	Dark
19	Circle	Red	Small	Dim	Fast	Light
20	Circle	Red	Large	Bright	Slow	Dark
21	Circle	Green	Small	Bright	Fast	Dark
22	Circle	Green	Large	Dim	Slow	Light
23	Circle	Blue	Small	Dim	Slow	Dark
24	Circle	Blue	Large	Bright	Fast	Light
25	Square	Red	Small	Dim	Fast	Light
26	Square	Red	Large	Bright	Slow	Dark
27	Square	Green	Small	Bright	Fast	Dark
28	Square	Green	Large	Dim	Slow	Light
29	Square	Blue	Small	Dim	Slow	Dark

It is essential to also model invalid or error cases.

The model designed so far is good to verify that the SUT does what it is supposed to do, that is, it correctly classifies the figures put in front of the camera, but what about not doing what it is not supposed to do? What happens if the figure does not have one of the defined shapes or colors? Does the system stop? Does it assign the same classification as to the last one processed? Does it classify it as unrecognizable and send it to a trash bin?

TABLE 5.4 Hybrid Test Strategy

Test Case Number	Triangle Type	Color	Size	Lighting	Speed	Background
1	Equilateral	Red	Small	Dim	Fast	Light
2	Scalene	Green	Large	Bright	Slow	Dark
3	Isosceles	Blue	*	*	*	*
Test Case Number	**Shape**	**Color**	**Size**	**Lighting**	**Speed**	**Background**
4	Circle	Red	Small	Dim	Fast	Light
5	Circle	Red	Large	Bright	Slow	Dark
6	Circle	Green	Small	Bright	Fast	Dark
7	Circle	Green	Large	Dim	Slow	Light
8	Circle	Blue	Small	Dim	Slow	Dark
9	Circle	Blue	Large	Bright	Fast	Light
10	Square	Red	Small	Bright	Slow	Light
11	Square	Red	Large	Dim	Fast	Dark
12	Square	Green	Small	Dim	Fast	Light
13	Square	Green	Large	Bright	Slow	Dark
14	Square	Blue	Small	Bright	Fast	Dark
15	Square	Blue	Large	Dim	Slow	Light
16	Triangle	Red	Small	Dim	Slow	Dark
17	Triangle	Red	Large	Bright	Fast	Light
18	Triangle	Green	Small	Bright	Slow	Light
19	Triangle	Green	Large	Dim	Fast	Dark
20	Triangle	Blue	Small	Dim	Fast	Light
21	Triangle	Blue	Large	Bright	Slow	Dark

Note: First, we establish the equivalence of processing for the different kinds of triangles. Second, we test for interaction effects using three-way combinatorial testing and triangle as an abstract parameter value that will be replaced with one of the actual types: isoceles, equilateral, or scalene during testing.

* Stands for any valid value.

Again, this might or might not have been specified, but from a black box perspective, it is necessary to test both perspectives: what does the system do when presented with a valid combination of inputs and what does it do when presented with an invalid one. The challenge with negative testing is that the inclusion of an invalid value in a test case might result in the premature termination of the processing, for example, with the display of an error message, and the subsequent discarding of valuable combinations of valid values included in it. This is the problem of ineffectual combinations to which we referred earlier. In practical terms, this means that although the generated test suite was of strength t, it is possible that not all the

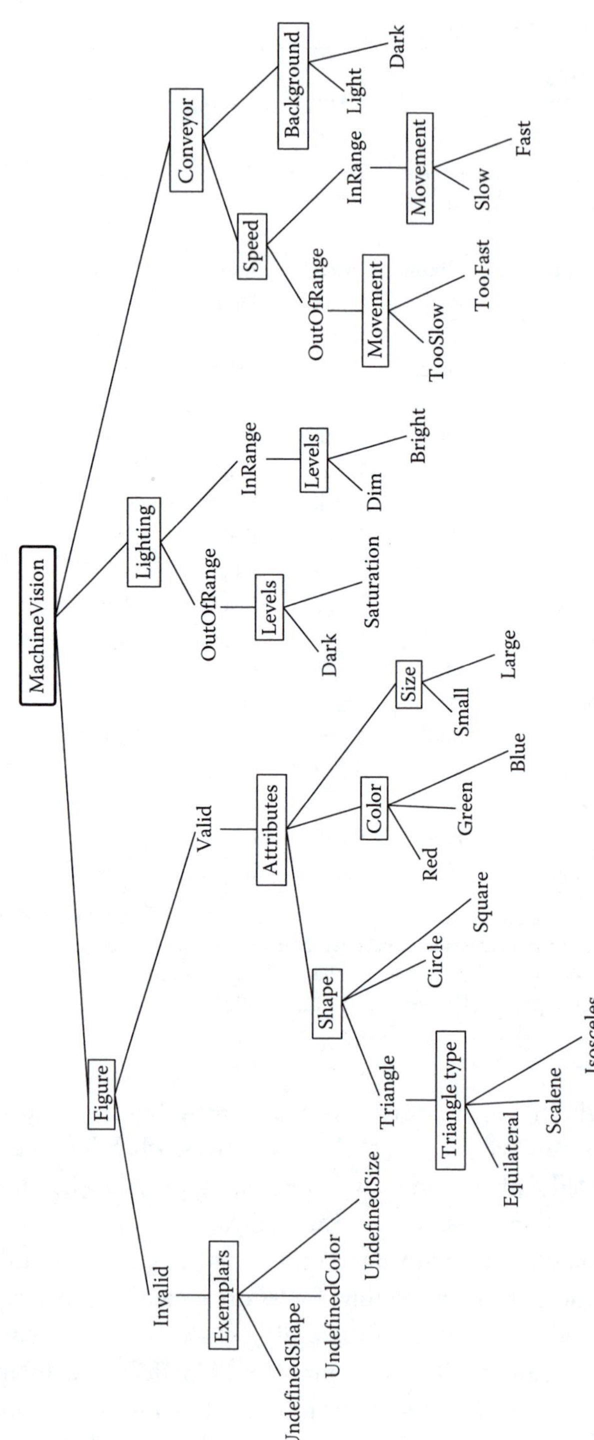

FIGURE 5.8 Classification tree showing separate branches for valid and invalid scenarios.

TABLE 5.5 All Single Values for Negative Testing

Test Case Number	Figure	Lighting	Speed	Background
1	UndefinedShape	Dim	Fast	Light
2	UndefinedColor	Bright	Slow	Dark
3	UndefinedSize	*	*	*
4	Triangle + red + small	Dark	*	*
5	Circle + green + large	Saturated	*	*
6	Square + blue + *	*	TooSlow	*
7	*	*	TooFast	*

* Stands for any valid value.

combinations of that order will be processed. Figure 5.8 illustrates a solution to the problem, which consists in the creation of two separate branches for each test parameter: one for valid values and the other for invalid ones. By separating the two branches, it is very simple to generate test cases using a test strategy, which consists of all singles for negative testing (see Table 5.5) and *t*-way combinations for all valid values from a single tree.

Although the trees in Figures 5.7 and 5.8 look quite similar, they are not. First, notice that "Figure," which was modeled as a composition, has become a classification with two possible scenarios: "Invalid" and "Valid." Second, we needed to introduce some artificial nodes like "Exemplars" and "Attributes" to comply with the CTM syntax. For all the other nodes, we follow the same strategy using "InRange" to denote normal operating conditions (valid values) and "OutOfRange" to describe abnormal ones (invalid values).

5.4.2 Audio Amplifier

In this example, the SUT is the audio amplifier shown in Figure 5.9. The amplifier has two input jacks, two volume controls, two toggle switches, and one three-way selection switch. The main purpose of this example is to

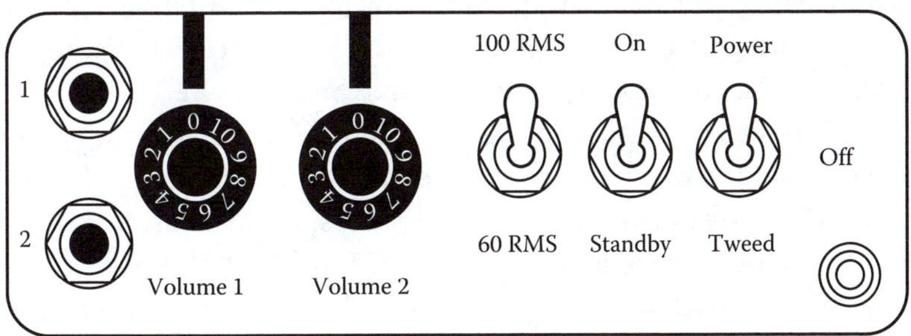

FIGURE 5.9 Audio amplifier example.

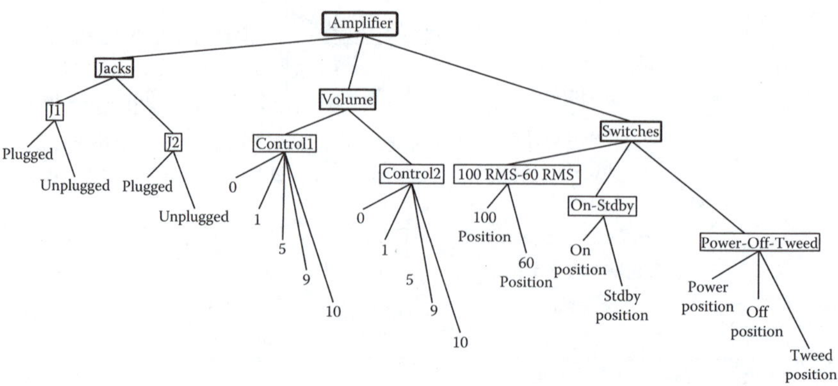

FIGURE 5.10 Classification tree for the audio amplifier including dominant values.

continue the discussion about ineffectual combinations briefly addressed in the previous example when discussing how to deal with invalid values. Specifically, we are addressing here the problem of dependent parameters. This problem arises when the relevance of one or more test parameters depends on the value of another parameter called the dominant parameter.

Figure 5.10 shows a straightforward tree for the SUT. The test-relevant aspects are whether the input jacks are plugged in or not, and the positions of the volume control and the three switches. While there is nothing wrong with the tree as drawn, if we were to mechanically map its nodes into the parameters and values of a combination table, the resulting test suite would provide a coverage lower than the table's nominal strength. To see how this is possible, it suffices to look at Table 5.6. When the amplifier is "Off" or in "Stdby," it is not amplifying and in consequence nothing else is testable. In terms of Table 5.6, this means that all value combinations included in a row that contains an "Off" or a "Stdby" are lost. In this particular instance, the problem could be easily resolved by physically removing the offending parameters values, or in the case of ACTS by applying the constraints: *PowerOffTweed* ≠ "Off" and *Stdby* = "On."

If one would have wanted to highlight the different modes the amplifier can be in and the corresponding switch positions, a tree like the one in Figure 5.11 can be drawn. In this case, we will split the test suite into three test groups, one for each mode as shown in Table 5.7.

Another approach would be to introduce dummy "Not Applicable" values, but in general this is not a good idea as it requires the introduction of constraints to eliminate them from where they do not belong.

TABLE 5.6 Test Suite Generated by the Tree in Figure 5.10 Highlighting Inefectual Combinations

Test Case Number	J1	J2	Control1	Control2	RMS	OnStdby	PowerOffTweed	Comment
1	Unplugged	Unplugged	0	0	60 Position	Stdby	Off	Ineffectual
2	Plugged	Plugged	0	1	100 Position	On	Power	
3	Unplugged	Plugged	0	5	60 Position	On	Tweed	
4	Unplugged	Unplugged	0	9	100 Position	Stdby	Power	Ineffectual
5	Plugged	Unplugged	0	10	60 Position	Stdby	Tweed	Ineffectual
6	Plugged	Plugged	1	0	100 Position	Stdby	Tweed	Ineffectual
7	Unplugged	Unplugged	1	1	60 Position	On	Off	Ineffectual
8	Plugged	Unplugged	1	5	60 Position	Stdby	Power	Ineffectual
9	Plugged	Plugged	1	9	100 Position	On	Off	Ineffectual
10	Unplugged	Plugged	1	10	100 Position	On	Power	
11	Plugged	Unplugged	5	0	60 Position	On	Power	
12	Unplugged	Plugged	5	1	100 Position	Stdby	Tweed	Ineffectual
13	Unplugged	Unplugged	5	5	100 Position	On	Off	Ineffectual
14	Unplugged	Plugged	5	9	60 Position	On	Tweed	
15	Plugged	Unplugged	5	10	60 Position	Stdby	Off	Ineffectual
16	Unplugged	Plugged	9	0	60 Position	Stdby	Tweed	Ineffectual
17	Plugged	Unplugged	9	1	100 Position	On	Off	Ineffectual
18	Unplugged	Plugged	9	5	60 Position	Stdby	Power	Ineffectual
19	Unplugged	Plugged	9	9	60 Position	On	Power	
20	Unplugged	Unplugged	9	10	60 Position	On	Off	Ineffectual
21	Unplugged	Plugged	10	0	60 Position	Stdby	Power	Ineffectual
22	Plugged	Unplugged	10	1	100 Position	On	Off	Ineffectual
23	Plugged	Unplugged	10	5	100 Position	Stdby	Tweed	Ineffectual
24	Unplugged	Plugged	10	9	100 Position	Stdby	Off	Ineffectual
25	Plugged	Plugged	10	10	100 Position	Stdby	Off	Ineffectual

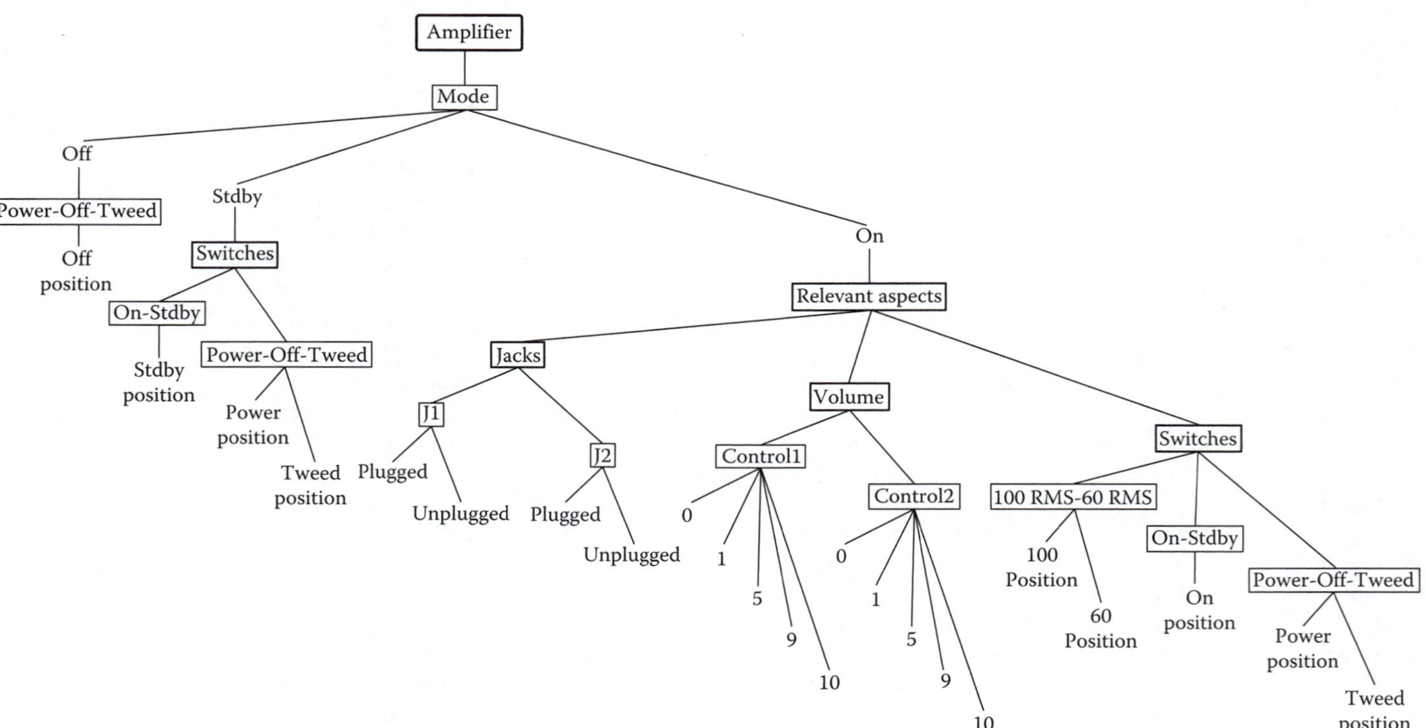

FIGURE 5.11 Classification tree for the audio amplifier highlighting the amplifier states and subordinating the dependent parameters to the mode in which they are relevant.

TABLE 5.7 Solution to the Combination Anomalies Generated by Introducing the Concept of "Mode"

Mode	Test Case Number	J1	J2	Control1	Control2	RMS	OnStdby	PowerOffTweed
Off	1							Off
Stdby	2						Stdby	Power
Stdby	3						Stdby	Tweed
On	4	Unplugged	Unplugged	0	0	60 Position	On	Tweed
On	5	Plugged	Plugged	0	1	100 Position	On	Power
On	6	Unplugged	Unplugged	0	5	100 Position	On	Power
On	7	Plugged	Unplugged	0	9	100 Position	On	Tweed
On	8	Unplugged	Plugged	0	10	60 Position	On	Tweed
On	9	Plugged	Plugged	1	0	60 Position	On	Power
On	10	Unplugged	Unplugged	1	1	100 Position	On	Tweed
On	11	Plugged	Plugged	1	5	60 Position	On	Tweed
On	12	Unplugged	Plugged	1	9	60 Position	On	Power
On	13	Plugged	Unplugged	1	10	100 Position	On	Power
On	14	Unplugged	Plugged	5	0	100 Position	On	Tweed
On	15	Plugged	Unplugged	5	1	60 Position	On	Power
On	16	Plugged	Unplugged	5	5	60 Position	On	Tweed
On	17	Plugged	Unplugged	5	9	100 Position	On	Tweed
On	18	Unplugged	Unplugged	5	10	60 Position	On	Power
On	19	Plugged	Unplugged	9	0	60 Position	On	Tweed
On	20	Unplugged	Plugged	9	1	100 Position	On	Power
On	21	Plugged	Unplugged	9	5	60 Position	On	Power
On	22	Plugged	Unplugged	9	9	100 Position	On	Tweed
On	23	Plugged	Plugged	9	10	100 Position	On	Power
On	24	Plugged	Unplugged	10	0	60 Position	On	Power
On	25	Unplugged	Plugged	10	1	100 Position	On	Tweed
On	26	Plugged	Unplugged	10	5	60 Position	On	Power
On	27	Plugged	Unplugged	10	9	60 Position	On	Power

5.4.3 Password Diagnoser

The password diagnoser is a software module executed as part of the registration process for an online banking system. The purpose of the diagnoser is to verify that user passwords conform to good security practices as defined in Table 5.8. The software analyzes the password submitted by a registrant and issues a diagnostic message for each requirement not met. This example highlights the distinction between using combinatorial testing to combine actual values as we did in the preceding example versus the combination of characteristics from which the actual test values are to be derived. The example also introduces the problems of missing combinations and infeasible combinations, and continues the discussion about ineffectual combinations.

In a traditional testing setting, the test designer might start by writing a series of passwords such as "Rick@992," "Eduardo4$," "Rick@1234," and "Eduardo," some of which will comply with the requirements and some of which will not. The problem with this approach is that with each password created, it becomes increasingly difficult to find a new one that tests an untried combination, increasing the risk of missing important combinations of characteristics.

One could perhaps concoct a tree like the one shown in Figure 5.12, where actual passwords can be generated by concatenating all combinations of the parameter values, but the reader will agree that it is not an elegant solution nor one that is general enough to use in all situations. Contrast this with the tree in Figure 5.13. In that tree, instead of using tokens corresponding to actual sequences of characters to be included in the password, we used the requirements the password must conform to as the values of the test parameters. The intention of the test designer as represented by the tree is undoubtedly more explicit than in the case of the tree in Figure 5.12. In this case, we will use combinatorial testing to generate specifications with which the actual test values must comply and then use the specifications to generate the necessary test cases manually or

TABLE 5.8 Password Requirements

Password Characteristic	Requirement
Length	Shall be eight characters or more
Composition	Shall include at least one uppercase character, one numeric character, and one special character
Predictability	Shall not have any QWERTY keyboard or ASCII sequence of length greater than 3

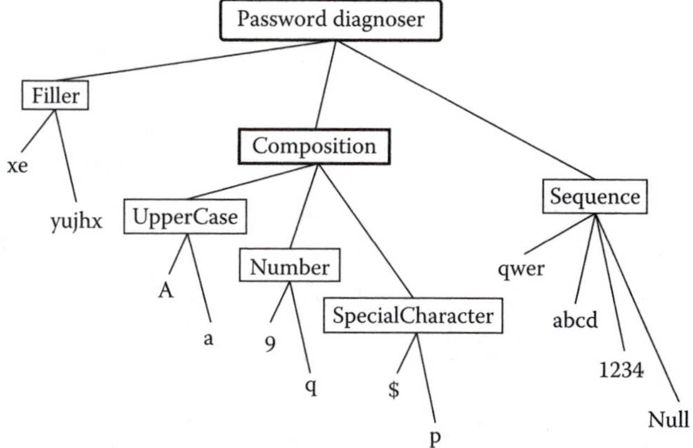

FIGURE 5.12 A value-based tree capable of generating test passwords through the concatenation of leaf values.

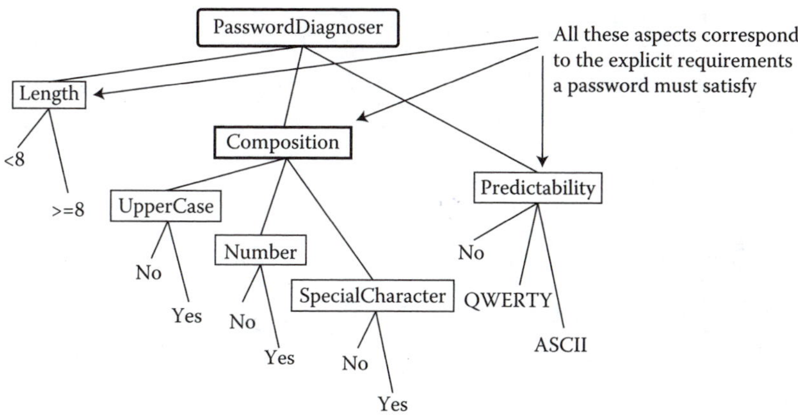

FIGURE 5.13 A characteristic-based tree making explicit the test designer intentions.

automatically. The result of the three-way combination of all test parameters is shown in Table 5.9.

A quick analysis of the entries in Table 5.9 shows that no valid password was generated. Without doubt, such a password must be part of any test suite that we could call adequate. A valid password must comply with five requirements: its length should be eight characters or more ("Length ≥ 8"), it should include at least one uppercase letter ("UpperCase = Yes"), a number ("Number = Yes"), and a special character ("SpecialCharacter = Yes"), and it must not include a sequence longer than

TABLE 5.9 Three-Way Interactions

Test Case Number	Test Specification	Test Value[a]	Expected Result (Diagnostic Message)	Actual Result
1	Length: >=8 UpperCase: Yes Number: Yes SpecialCharacter: No Predictability: QWERTY	Eduardo7890	Password must include at least one special character Password must not contain more than three consecutive keyboard keys	
2	Length: >=8 UpperCase: No Number: No SpecialCharacter: No Predictability: No	eduardo974	Password must include at least one uppercase letter Password must include at least a number Password must include at least one special character	
3	Length: >=8 UpperCase: Yes Number: No SpecialCharacter: Yes Predictability: ASCII	Eduardo#abcd	Password must include at least a number Password must not contain more than three consecutive letters or numbers	
4	Length: <8 UpperCase: Yes Number: Yes SpecialCharacter: Yes Predictability: No	Rick$1	Password must have at least eight characters	
5	Length: <8 UpperCase: No Number: No SpecialCharacter: Yes Predictability: QWERTY	xcvb$a	Password must have at least eight characters Password must include at least one uppercase letter Password must include at least a number Password must not contain more than three consecutive keyboard keys	

6	Length: <8 UpperCase: No Number: Yes SpecialCharacter: No Predictability: ASCII	arstu4	Password must have at least eight characters Password must include at least one uppercase letter Password must include at least one special character Password must not contain more than three consecutive letters or numbers
7	Length: <8 UpperCase: Yes Number: No SpecialCharacter: No Predictability: ASCII	Peopqr	Password must have at least eight characters Password must include at least a number Password must include at least one special character Password must not contain more than three consecutive letters or numbers
8	Length: >=8 UpperCase: No Number: Yes SpecialCharacter: Yes Predictability: No	hotspot2*	Password must include at least one uppercase letter
9	Length: <8 UpperCase: Yes Number: No SpecialCharacter: No Predictability: QWERTY	Asdf	Password must have at least eight characters Password must include at least a number Password must include at least one special character Password must not contain more than three consecutive keyboard keys
	. . .		
18	Length: >=8 UpperCase: No Number: No SpecialCharacter: No Predictability: ASCII	eduardocdef	Password must include at least one uppercase letter Password must include at least a number Password must include at least one special character Password must not contain more than three consecutive letters or numbers

[a] These values are written by the test designer following the specification.

three characters ("Sequence = No"), but because we chose to generate a test suite consisting of all three-way combinations, there is no guarantee that such a combination will be included. This is what we have called the problem of missing combinations. This problem can be addressed by "seeding" [50] handcrafted test cases to cover specific combinations, such as a "valid" password, or with a tool such as ACTS, by using variable strength combinatorial testing and applying all combinations to the required test parameters to be sure that mandatory combinations are not missed (see Appendix D). The second solution is only advisable when the required combinations are made up of small subset of test parameters as otherwise we will be forcing a large number of test cases just to cover a few required combinations.

If you have not noticed, it seems that the test designer has a tendency to capitalize the first letter of the passwords and to write the sequences in ascending or left to right order, and one cannot but wonder whether the developer based his design on the same thought patterns: did the developer test for the presence of an uppercase letter in any position other than the beginning of the string? Did he consider descending or right to left sequences? To prevent these biases, the test designer might want to include other test-relevant aspects such as the position where the characters appear and the sequence order (see Figure 5.14). Notice that these new test parameters do not come from the diagnoser specification but from the experience of the test designer or from organizational assets such as test catalogs.

Putting aside for the time being whether it is a good strategy or not, if we were to combine all the leaves of the tree in Figure 5.14, we would produce a large number of test cases, including many infeasible sequences; see, for example, test cases 21, 22, and 69 in Table 5.10, where an "UpperCase" letter and a "Number" occupy the same position in a string, a situation which is physically impossible. Test case 70 is valid since the "InBetween" position is not a single position but any position between the beginning and the end of the string. We can deal with the infeasible test cases in different ways. We can try to prevent the generation of infeasible test cases by modeling the problem in a different way; see Figure 5.15 in which we model "Position" as a classification to which we subordinate the character types: "LowerCase," "UpperCase," "Number," and "SpecialCharacter." Notice that from the point of view of the intellectual effort required to create a tree, it is better to start treating every test parameter independently from each other like we did for the tree in Figure 5.14 and then rearrange it, than trying to figure the "optimum" tree from the beginning. With regard to the "Predictability" test parameter, if we were to "flatten–out" the tree

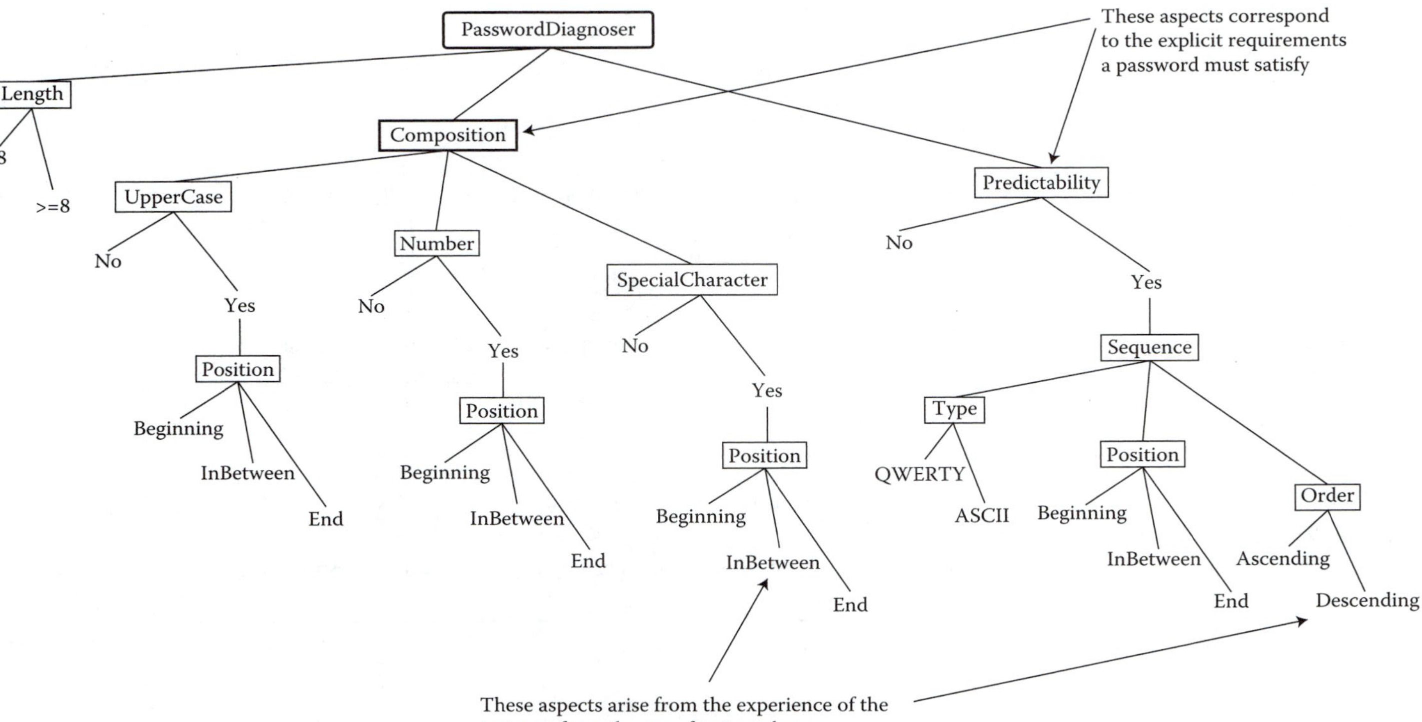

FIGURE 5.14 A refined classification tree including implementation aspects.

TABLE 5.10 Infeasible Test Cases

Test Case Number	Length	UpperCase	Number	SpecialCharacter	SequenceType	SequencePosition	SequenceOrder	Test value
1	<8	No	No	No	No	Beginning	Ascending	Ineffectual
2	>=8	No	No	Beginning	QWERTY	InBetween	Descending	@lkjheduardo
				. . .				
21	<8	Beginning	Beginning	No	No	InBetween	Descending	Infeasible
22	<8	Beginning	Beginning	Beginning	QWERTY	Beginning	Ascending	Infeasible
				. . .				
69	>=8	End	End	No	No	InBetween	Descending	Infeasible
70	>=8	Beginning	InBetween	InBetween	QWERTY	End	Descending	E9#eduardolkjh

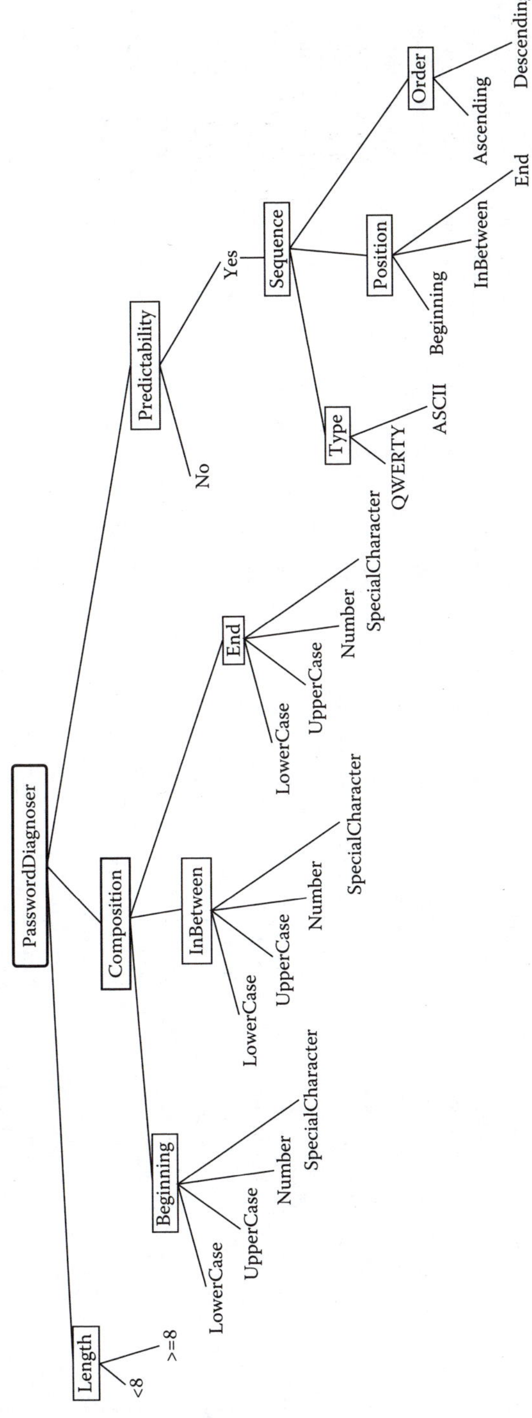

FIGURE 5.15 This model of the input domain for the password diagnoser produces less infeasible test cases than the one in Figure 5.14.

TABLE 5.11 Two-Way Interactions for the Password Diagnoser Example

Test Case					Predictability			
Number	Length	Beginning	InBetween	End	Type	Position	Order	Comment
1	≥8	UpperCase	Number	SpecialCharacter	QWERTY	Beginning	Descending	Predictability testing
2	≥8	UpperCase	Number	SpecialCharacter	ASCII	Beginning	Ascending	Predictability testing
3	≥8	UpperCase	Number	SpecialCharacter	QWERTY	InBetween	Ascending	Predictability testing
4	≥8	UpperCase	Number	SpecialCharacter	ASCII	InBetween	Descending	Predictability testing
5	≥8	UpperCase	Number	SpecialCharacter	QWERTY	End	Ascending	Predictability testing
6	≥8	UpperCase	Number	SpecialCharacter	ASCII	End	Descending	Predictability testing
7	≥8	LowerCase	LowerCase	UpperCase	Yes[a]			Diagnostic testing
8	<8	LowerCase	UpperCase	SpecialCharacter	No			Diagnostic testing
9	≥8	LowerCase	Number	Number	No			Diagnostic testing
10	≥8	LowerCase	SpecialCharacter	LowerCase	No			Diagnostic testing
11	<8	UpperCase	LowerCase	Number	Yes			Diagnostic testing
12	≥8	UpperCase	UpperCase	LowerCase	Yes			Diagnostic testing
13	<8	UpperCase	Number	SpecialCharacter	Yes			Diagnostic testing
14	<8	UpperCase	SpecialCharacter	UpperCase	No			Diagnostic testing
15	≥8	Number	LowerCase	SpecialCharacter	No			Diagnostic testing
16	<8	Number	UpperCase	UpperCase	Yes			Diagnostic testing
17	<8	Number	Number	LowerCase	Yes			Diagnostic testing
18	<8	Number	SpecialCharacter	Number	Yes			Diagnostic testing
19	≥8	SpecialCharacter	LowerCase	LowerCase	No			Diagnostic testing
20	<8	SpecialCharacter	UpperCase	Number	Yes			Diagnostic testing
21	≥8	SpecialCharacter	Number	UpperCase	No			Diagnostic testing
22	≥8	SpecialCharacter	SpecialCharacter	SpecialCharacter	Yes			Diagnostic testing
23[b]	≥8	Number	SpecialCharacter	Number	No			Seeded test case

[a] Any sequence will suffice since their equivalence was verified by the predictability testing suite (test cases 1 through 6).

[b] This test case was seeded.

and include all the test parameters in the same combination table, we will be generating a lot of ineffectual combinations due to the fact that the "No" value of the "Predictability" parameter would become a dominant value controlling whether "Type," "Position," and "Order" parameters are relevant or not. This is a problem akin to what we face in the design of a relational database in which we must break down a hierarchical structure into several tables to avoid update anomalies. To solve the problem, we will split the test suite into three test groups as shown in Table 5.11. The first test group will test for "Sequence" with different characteristics assuming all other requirements are met. The second group tests for interaction effects between all requirements assuming that all variations of "Sequence" form an equivalence class. The third group defines a password that satisfies all requirements since there is no guarantee that such a test case would be generated using two-way interactions.

5.5 SELECTING THE SYSTEM UNDER TEST

The first step in the modeling process is to select the SUT and define the scope of testing. This apparently simple choice seems to be taken for granted in the literature, but it has a large effect in the complexity of the resulting models and the extent to which we will experience combination anomalies. To see why, let us look at the "Replace" form example in Figure 5.16. The "Replace" form encompasses multiple functionalities: "FindNext," "Replace," "ReplaceAll," "Help," and "Close."

If we were to choose the SUT to be the "Replace" form, we could end up with a tree like the one shown in Figure 5.17. The problem with this tree is that the "Command" test parameter has many dominant values, for example, "Help," "Close," and "Find Next," which will result in a large number of ineffectual combinations. A better modeling approach would be to partition the input domain according to "Commands" (see Figure 5.18a). The

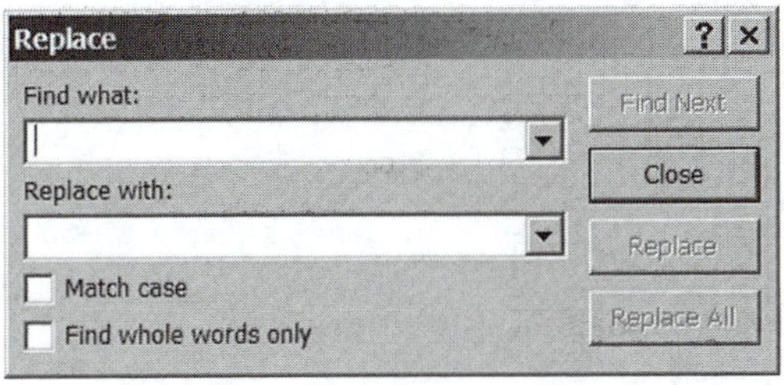

FIGURE 5.16 Replace form. One or multiple SUTs?

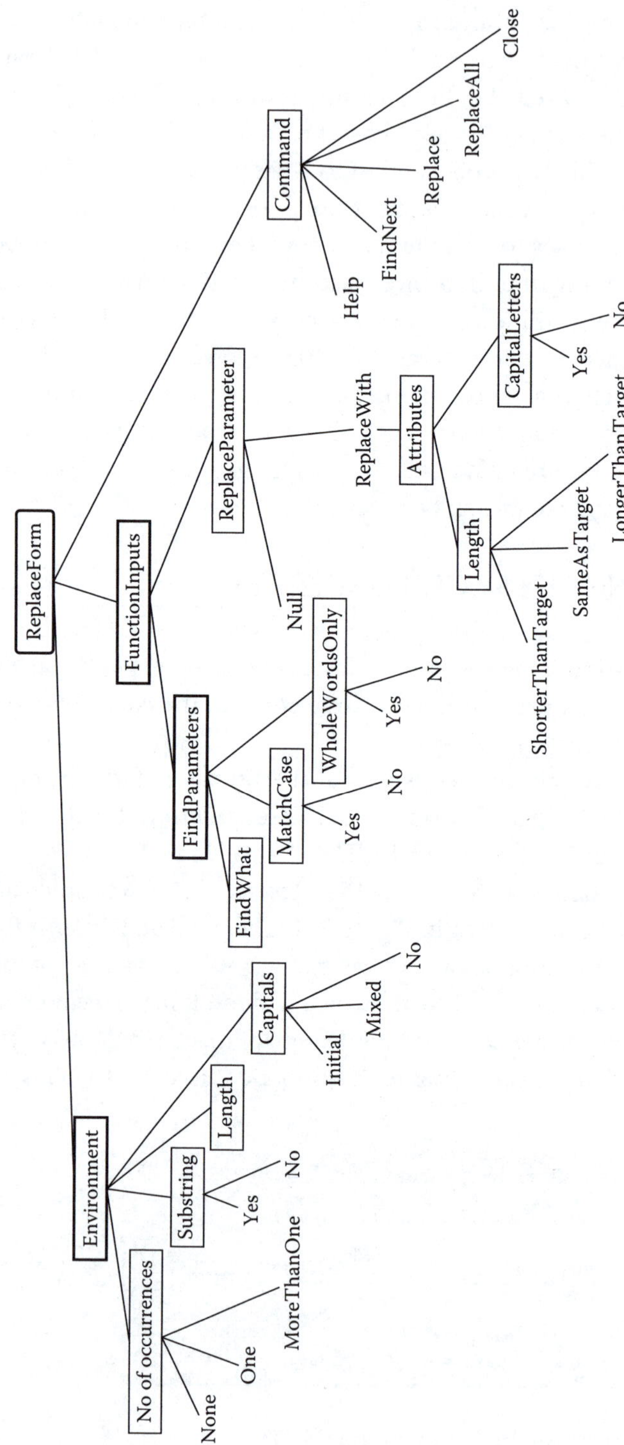

FIGURE 5.17 The scope of testing is the Replace form and, in consequence, the particular command to be executed becomes a test parameter.

only problem with this tree is that the current generation of tools does not support the reuse of subtrees and in consequence the model will be huge. We expect this limitation to be addressed by tool suppliers as the method gains acceptance. Figure 5.18b shows a model where the test parameters were hierarchically organized, from the most common to the least common according to the command type. This solution required the introduction of artificial nodes, which diminish the understandability of the model. Figure 5.18c shows a separate tree for each command. The problem with this solution is the high cost of maintenance and the possibility of inconsistencies over time and in case the trees are developed by different test designers.

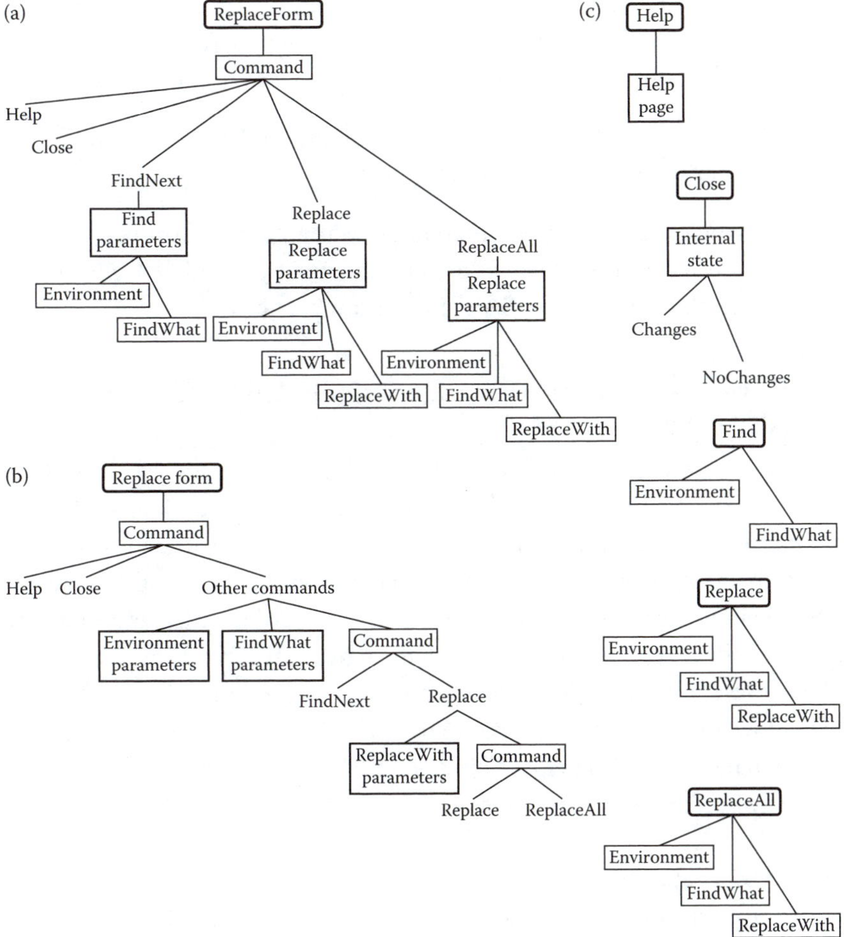

FIGURE 5.18 Modeling the "Replace" form. Single versus multiple SUT strategy. (a) One subtree for each command. (b) Common parameters. (c) Multiple trees.

Besides "you'll know when you see it," there are a few guidelines that can be followed when choosing the SUT:

- A well-defined function can be usually named with a verb and a noun. The verb defines the action required and the noun describes what is acted upon. If you cannot name the SUT this way, it is probably too big and embeds many functions or it is too small and only realizes part of a function.
- SUT execution must be initiated by a specific event consistent with the level of testing. If you are performing unit testing, the SUT execution must be initiated by a method, procedure, subroutine, or similar type of call. If you are performing system testing, the trigger must be a GUI command, a condition, or a temporal event known to the user.
- The result must be discrete and identifiable. That is, you can differentiate individual instances of the result.
- Mixing functionality of different arity and parameter types must be avoided. The consequences of not following this guideline were illustrated through the "Replace" form example above.

5.6 COMBINATORIAL TESTING AND BOUNDARY VALUE ANALYSIS

BVA is a widely used black box testing technique that consists of testing the SUT with values on and off an equivalence class boundary under the assumption that these are bug-prone regions. The aim is to detect domain faults, that is, applying a correct function to the wrong data. In this analysis, we will circumscribe to the most common use of BVA, which assumes that each boundary segment or border is defined by a linear predicate on a single variable forming a convex set (see Figure 5.19). These properties are the ones that allow for each test parameter to be varied independently from the others. Each border is either closed or open. A domain boundary is closed with respect to a domain if the points on the boundary belong to the domain and open if the points in it belong to another domain. A closed border is formed by a value and a relational expression with a $\leq$, $\geq$, or $=$ operator. An open border results from a value and a relational expression with a $<$, $>$, or $\neq$ operator. These restrictions are not restrictions in the sense that you cannot use the techniques described if they are violated, but in the sense that if you apply them to such cases, tests are unlikely to be productive [13].

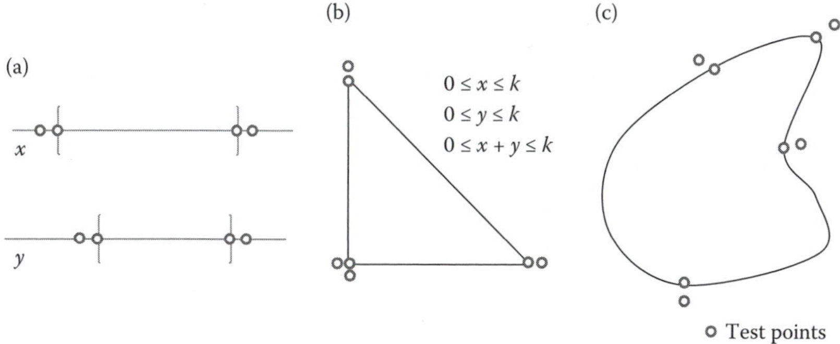

FIGURE 5.19 Nicest (a), nice (b), and ugly domains (c). The same number of test cases does not provide the same level of confidence in all cases. (a) Both variables are independent. (b) Functional linear relationship between x and y. Convex domain. (c) Nonlinear relationship between x and y. Concave domain.

Other forms of BVA include domain testing [46,217], which looks at the execution paths to derive the partition of the SUT input domain and allows linear relationships among the domain variables.

The basic question we are trying to answer here is whether we should use BVA values in interaction testing or first establish confidence that the partition is indeed an equivalence class and there are no boundary problems and then test for interactions using a nominal value from it.

While a definite answer to this question has proven elusive without making assumptions to the point of making it too restrictive for any practical use, we can analyze the issues and trade-offs for specific situations from two perspectives: the number of test cases required and what we get from them. For this, we will use the example in Figure 5.20, which describes a tree for the input domain defined by $EC_1 = \{(x, y) | 2 \le x \le 10 \text{ and } 1 \le y \le 7\}$ and $EC_2 = \{(x, y) | 2 \le x \le 10 \text{ and } 7 \le y \le 12\}$ to illustrate the process. Figure 5.21 shows the input domain and the corresponding test cases for the four strategies outlined below and Figure 5.22 shows the number of test cases required as a function of the number of test parameters.

The analyzed testing strategies are

- BVA once each. In this strategy, we just make sure that each BVA value is included at least once in the test suite.
- BVA once each + All pairs EC. This strategy includes the above-described BVA once each followed by all possible pair of values at the equivalence class level.

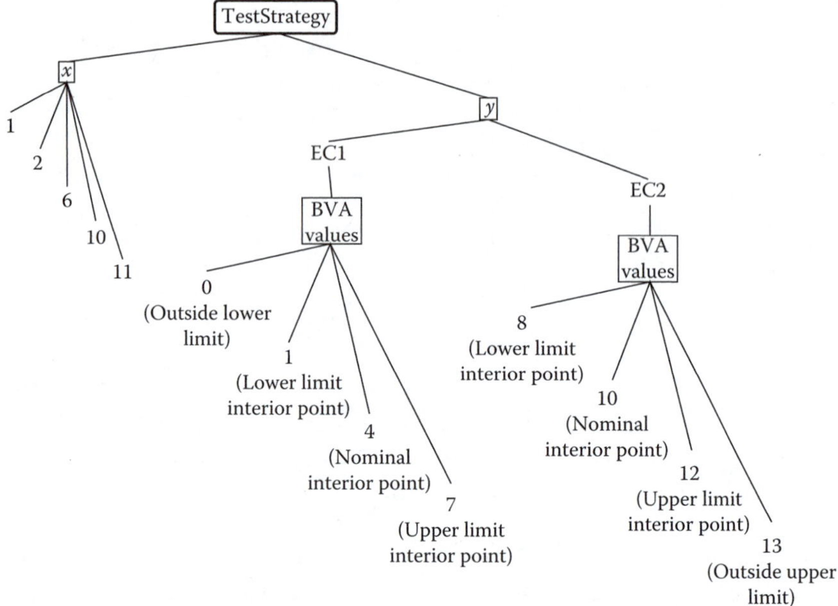

FIGURE 5.20 Should we test first for all values under each equivalence class and then use interaction testing at the EC1 and EC2 level or should we use combinatorial testing with all the values generated by BVA?

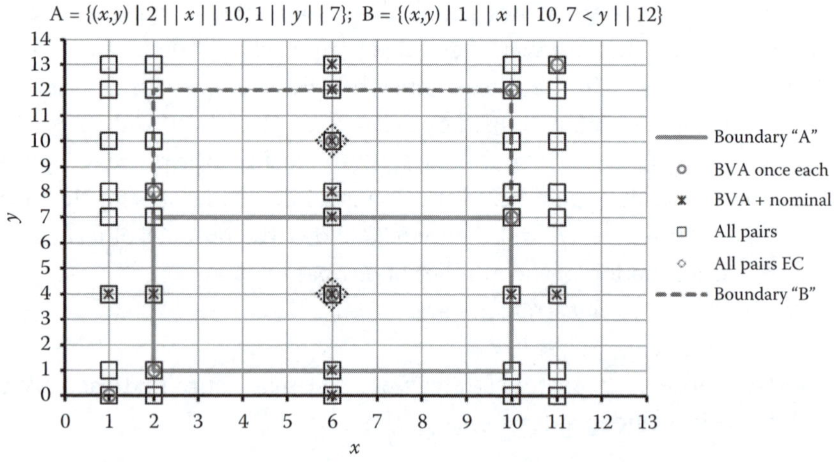

FIGURE 5.21 **(See color insert.)** A two-dimensional convex region tested using different strategies.

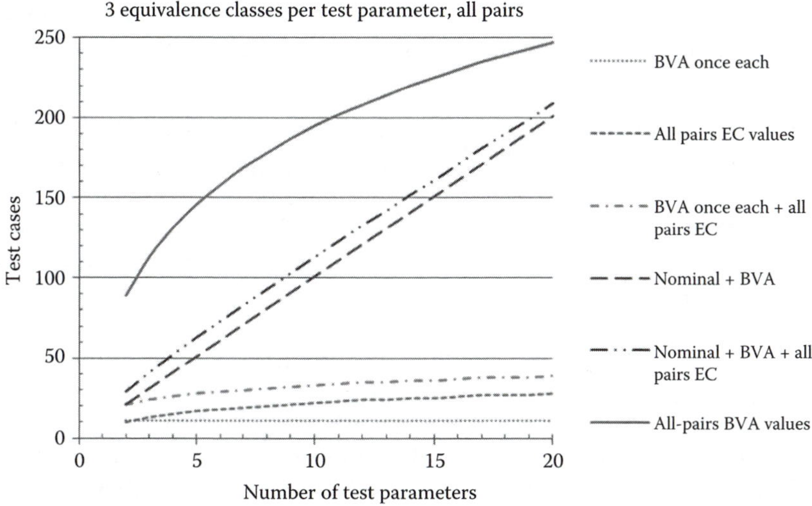

FIGURE 5.22 **(See color insert.)** Number of test cases required by different testing strategies.

- BVA + Nominal. In this case, we combine each of the BVA values from each parameter with a nominal value from all the other parameters.
- All pairs. All possible pairs of BVA values are included in at least one test.

Testing all BVA values for two-way interactions in the example requires 40 test cases. Testing first for all equivalences using each value once followed by interaction testing will require 8 + 2 = 10 test cases.

Other things being equal, for 10 test parameters, the number of test cases required by the "BVA once each + All pairs EC" is a sixth of that required by the "All pairs BVA": 33 versus 195. A more thorough strategy, "Nominal + All BVA," which tests all BVA values from each parameter with a single nominal value from each of the other test parameters, followed by "All pairs of EC" values requires little more than half the values of "All pairs," 113 versus 195, but four times more than the "BVA once each + All pairs EC strategy." In this example, from the point of view of the economy of test cases, the best strategy would be to first establish the equivalence of each class and then test for interactions using a nominal value. But what are we giving up when we choose the most economical alternative?

Boundary errors refer to the manner in which the program processes values at boundaries, and some types of errors include:

- *Closure:* A boundary is closed or open when it should have been the opposite, such as when ≥ is used when > is the correct operator.
- *Missing boundary:* Distinction between values is not processed, such as when a conditional is missing that should be present to distinguish between two classes of input.
- *Parallel boundary shift:* Boundary is uniformly shifted across values, such as `x` > `20`, when `x` > `10` is the correct condition.
- *Boundary tilt:* Boundary is shifted because coefficients in a condition are wrong, such as `y` > `2*x`, where the correct condition is `y` > `x` (picture how these would be seen on a graph).

The BVA once each strategy can consistently detect closure, missing boundaries, and border shifts for parallel boundaries but cannot be relied upon to detect boundary tilts or spurious border problems. Because of the combinations of two or more "off limits" values, input masking might occur at the extremes.

The BVA + Nominal strategy is also capable of consistently detecting closure, missing boundaries, and parallel shifts. The larger number of test points will contribute to expose boundary tilts and spurious boundaries but there is no certitude any or all will be found. The possibility of masking is removed by using only one off-region value at a time.

Similar to the other two, the all-pairs strategy will consistently detect closure, missing boundaries, and parallel shifts. It will also detect spurious relationships among variables of rectangular domains by testing all corners of the domain. The possibility of masking exists because there are test cases that include two out-of-limits values but this is compensated by the combination of the off-limit values with the nominal ones.

Based on the above analysis, the test designer could design, for example, a test strategy like the one shown in Figure 5.23 in which "off limits" values from each parameter are combined with all EC values from the other parameter to test boundaries without risk of input masking due to the combination of two or more "off limits" values. The all-interior points establish equivalence of the proposed classes and combinatorial testing is applied at the EC level. Notice that in the present example, because of the

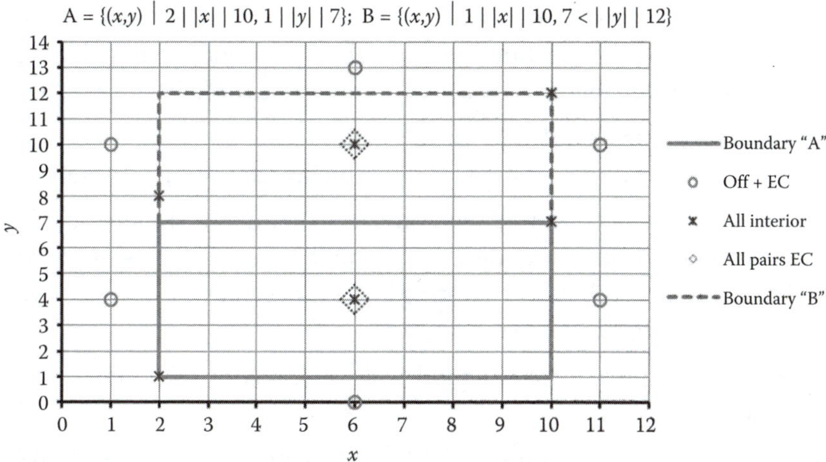

FIGURE 5.23 A hybrid testing strategy combining BVA with a all-pairs at the equivalence class level.

number of variables and values involved, the all-pairs test points coincide with some of the all-interior points.

5.7 CHAPTER SUMMARY

Identifying the SUT, the test parameters, and the parameter values is not a mechanical task. It requires knowledge and ingenuity. Attention must be paid to missing, infeasible, and ineffectual combinations. Failure to do this will result in test suites that provide a lower effective coverage than that used to build the combination tables at best, and are inadequate at worst.

- The CTM provides a systematic way to analyze and document the input domain of the selected SUT, but it is not replacement for thinking.
- To limit the number of test cases to what practicality warrants, we should employ hybrid strategies like verifying equivalence first and testing for interactions second.
- There are few bad tests—but a lot of unproductive ones.

REVIEW

Q: What factors should be considered part of the SUT's input domain for the purpose of testing?

A: The input domain is not restricted to the values of explicit inputs, such as variables in a procedure call or method invocation, but includes all data objects referenced by the SUT whose values or format are believed to have an effect on the their processing.

Q: Is the classification tree method a combinatorial testing technique?

A: No. The classification tree method is a structured technique for analyzing and documenting the input domain of an SUT. It is useful as a front end for combinatorial testing but is independent of the technique used to generate test cases.

Q: Is the classification tree method is a black box or white box testing technique?

A: The classification tree method advocates using all sources of knowledge to generate relevant test cases, whether functional specifications or design information.

Q: What are the three types of combinatorial anomalies commonly arising in practice?

A: Missing combinations, infeasible combinations, and ineffectual combinations. Missing combinations are specific combinations required by the test designer for which there is no guarantee that the combination algorithm will produce. Infeasible combinations are combinations that are generated by the combinatorial algorithm but are not physically or logically realizable in actual test cases. This situation occurs because the algorithms generating the combinations have no knowledge of the domain semantics. Ineffectual combinations have the same effect as infeasible combinations: loss of effective coverage. There are basically two scenarios: A value forces the brushing aside of other valuable combinations of parameter values included in the same test case that might never be tried. In other cases, certain test parameters are relevant only when a third parameter takes a given value but since all parameters are combined together irrespective of values, some of the combinations are wasted.

Q: Can entities—classes, classifications, and categories—at different levels of the tree hierarchy be used as test parameter values or should we only consider leaf classes?

A: Yes. Depending on the purpose of the test, any element in the hierarchy can be used as an abstract test value. Of course, for the purpose of executing the test, the abstract value will have to be replaced with an actual value or configuration.

CHAPTER 6

Managing System State

George Sherwood

A PERVASIVE CHALLENGE IN DESIGNING combinatorial test inputs is managing the state of the system under test (SUT). Different states may respond differently to inputs, so controlling the state is essential to knowing what is being tested and how thoroughly. When a set of test inputs leads to an unplanned result (e.g., initiation of a different feature or an exception) instead of the expected result—and if the SUT is not at fault—some of the *t*-tuples may not be covering the intended test area [63,179,194]. These test cases mask an expected system behavior. Consequently, gaps in the test coverage may leave some faults undetected. The need then is control of the system state from the set-up of the test case to its expected results. Attention to the state of the system can enable the tester to design a more complete plan for testing more states, to apply the test inputs more accurately with appropriate partitioning, and to be better informed about the inevitable limitations and tradeoffs of the test design.

This chapter introduces test models to apply combinatorial testing to the operation of complex systems. The models range from informal handling of SUT state, without requiring a complete specification, to test models based on transitions of state machines with one or more regions. The different models offer a range of tradeoffs for different development methods and goals. Describing system operation in terms of state machines can help meet the challenge of incorporating partitions with constraints into combinatorial test designs: The state machine clarifies system behaviors and helps to avoid test design errors and omissions. The examples given here illustrate a systematic procedure for generating test designs from state machines. The examples, models, and procedure are described in more detail in the Testcover.com tutorial [197]. These test designs illustrate pairwise coverage, but the technique is general and can be applied to higher strength *t*-way test designs as well.

State machines and methods for testing them have been studied and developed over many years [16]. In this chapter, the purpose of applying combinatorial testing to state machines is to complement existing techniques in situations where there are too many program variables (e.g., inputs) to cover all combinations of values. The test models depend on states of Unified Modeling Language (UML) state machine diagrams [82,151] as well as associated program variables. Together the diagrammed state and program variables define the *extended state*, which determines SUT behaviors and test results [169]. A key aim of the chapter is to show how test factor partitions can be defined to set up inputs that can reach each source state, to be consistent with the trigger and guard conditions, and to verify boundary values.

The chapter includes six sections. The first describes the need for and use of partitions with constraints in designing combinatorial test inputs. Four methods for simplifying constraints are presented in the second section. The third section introduces the *sequence unit replay model*, which uses informal management of the SUT state for testing an ecommerce shopping cart. The fourth section uses a state machine with one region to describe the same shopping cart. Its state machine leads to test designs based on the *transition model* and the *target state model*. The fifth section presents three test models spanning multiple orthogonal regions of a thermostat state machine, and the sixth section is a chapter review.

6.1 TEST FACTOR PARTITIONS WITH STATE

Equivalence partitioning is a well-known technique for separating values of a test input into classes of equivalent expected results. With *boundary value analysis,* a small number of values can be selected from each partition to exercise the SUT for the corresponding result, and to check that the partition boundaries are implemented correctly. Sections 4.1 and 5.6 described these methods as applied to input values. In this chapter, a *behavior* refers to a class of *equivalent results involving transition(s) among diagrammed states*. The specific results comprising the equivalence class reflect the program variables of the extended state as well. Thus, each partition has an expected behavior, and the partition's test cases initiate specific expected results.

An application for Fahrenheit to Celsius temperature conversion provides a simple example of partitioning. For inputs above absolute zero, the application is required to convert the temperature according to the usual formula; otherwise, an exception response is given. These two behaviors correspond to two partitions, and a small number of input values can be selected from each.

In more complex applications, there may be more classes of equivalent results. Additional partitions are needed then: Different normal behaviors require different partitions. And if different types of exception processing are to be verified individually, they need separate partitions as well. Moreover, the presence of multiple inputs implies partitions with multiple dimensions—one for each input—as in the following example.

6.1.1 Partitions for Expected Results

Figure 6.1 illustrates a design-your-burger application using an HTML form [181]. Inputs include a burger menu, a cooking choice radio button, eight topping checkboxes, buttons to reset or submit the form, and links to other pages. The purpose of the test is to verify the action of the form with all pairs of input values. A reasonable starting point is to associate each input with a test factor. However, four of the inputs—the reset, submit, and link inputs—are different from the others because they change the state. The submit button is the trigger to start the processing of the form's data.

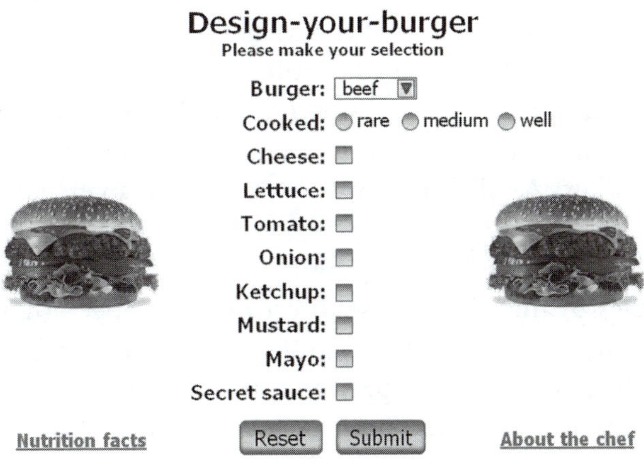

FIGURE 6.1 Design-your-burger form. Inputs include a burger menu (beef, turkey, or veggie), a cooking choice radio button, eight topping checkboxes, buttons to reset or submit the form and links to other pages.

The reset button clears the form, and the links go to the *Nutrition facts* page or the *About the chef* page. This partition is to verify the behavior of the submit button and the resulting processing. Inputs from the reset button and the links do not lead to the expected result state and should not be included in the partition. If all the inputs *are* put into the same partition, we can expect some input value pairs will not test the action of the form. Instead they will be reset or lost on the way to other pages.

Table 6.1 shows the 10 test factors chosen for this partition—the menu, radio button, and eight checkboxes—and their test cases. The submit button is clicked after the other input values are set. Each pair of input values is covered.

6.1.2 Partitions with Constraints

Typically test inputs need constraints so that the combinations of values in a partition lead to equivalent expected results (one behavior). If an interface takes months, days, and years as inputs for valid dates, day 31 should give a normal result only for the long months, January, March, May, and so on. If April 31 is a possible input, it should lead to an exception in a different partition instead. April 31 is an example of an invalid combination as described in Section 3.3.2. However, the reason that April 31 requires a different partition from April 30 is that a different behavior is expected from the SUT. When two *valid* inputs are expected to cause different behaviors, they need to be in separate partitions as well. Generally, input constraints are needed to ensure that each partition has a set of test cases that covers its equivalence class and does not encroach into another partition.

Combinations that belong in a partition are *allowed* by the partition; those that do not are *disallowed*. A partition with valid dates allows April 30 but not April 31. A partition with invalid dates that have too many days in a month allows April 31 but not April 30 (e.g., for an exception).

6.1.3 Direct Product Block Notation

The examples in this chapter use direct product block (DPB) notation [197] to specify test factor values and their constraints. In DPB notation each factor's allowed values appear on a separate line in a block, and all combinations of values in the block are allowed. *Partitions* consist of one or more *blocks*. A partition's allowed combinations are the union of its blocks' combinations. Importantly, the blocks are defined so that they do not include any disallowed combinations. Table 6.2 gives an example of DPB notation to specify valid dates according to calendar rules. All

TABLE 6.1 Design-Your-Burger Form Inputs

Burger	Cooked	Cheese	Lettuce	Tomato	Onion	Ketchup	Mustard	Mayo	Secret Sauce
Turkey	Well	Yes	Yes	No	Yes	No	No	No	Yes
Veggie	None	Yes	No	Yes	No	Yes	Yes	Yes	No
Beef	Rare	No	Yes	No	Yes	Yes	No	Yes	No
Turkey	Medium	No	No	Yes	Yes	Yes	Yes	Yes	Yes
Beef	None	No	No	No	No	No	No	No	Yes
Veggie	Rare	Yes	No	Yes	No	No	Yes	No	Yes
Veggie	Medium	Yes	Yes	No	No	No	No	Yes	No
Beef	Well	No	No	Yes	No	Yes	Yes	Yes	Yes
Turkey	Rare	No	Yes	Yes	No	Yes	No	No	No
Turkey	None	Yes	Yes	No	Yes	No	Yes	No	No
Veggie	Well	No	Yes	No	Yes	Yes	No	Yes	No
Beef	Medium	Yes	Yes	Yes	Yes	No	No	No	Yes

Source: Adapted from Sherwood, G. B. *Getting the Most from Pairwise Testing: A Guide for Practicing Software Engineers*. CreateSpace, Scotts Valley, California, 2011.

Note: This pairwise design tests the action of the submit button in Figure 6.1. All pairs of test inputs are covered in these 12 test cases. Inputs for the reset button and the links to other pages are excluded from this partition because they change the state without leading to the expected result.

TABLE 6.2 Partition for Valid Calendar Dates in DPB Notation

```
Calendar Example
Month
Day
Year
# valid dates partition
+ all months, first & tenth days
jan feb mar apr may jun jul aug sep oct nov dec
1 10
2015 2016 2017
+ long month last day
jan mar may jul aug oct dec
31
2015 2016 2017
+ short month last day
apr jun sep nov
30
2015 2016 2017
+ feb last day, common year
feb
28
2015 2017
+ feb last day, leap year
feb
29
2016
```

Source: Courtesy of Testcover.com. www.testcover.com.

Note: The first four lines give names for the request and the three test factors. The next line, starting with the pound (#), starts the partition for valid dates. The partition boundaries conform to calendar rules using five blocks, indicated by lines starting with plus (+) and a comment. The first block includes all combinations of months and selected years, with first and tenth days of the month. The second and third blocks split months according to length and use the corresponding last day (31 or 30). The fourth and fifth blocks split years into common and leap years for February's last day (28 or 29).

combinations from each block are allowed, and the set of all allowed combinations conforms to the partition boundaries.

DPB notation is independent of the strength of the design. That is, the same DPB partition defines the allowed t-tuples for any t up to the total number of factors. There can be various ways to choose the blocks in a partition. DPB notation allows blocks that are disjoint or overlapping in various ways. These block definitions are equivalent when they specify the same combinations. A typical approach for choosing a partition's blocks is to select a block that covers most or all of the partition, and then split and/or add blocks to account for dependencies and boundaries. The calendar example in Table 6.2 illustrates this approach as well.

6.2 TEST FACTOR SIMPLIFICATIONS

Complex systems can have complicated constraints, but some constraints lend themselves to simplification. This section describes four situations in which constraints can be simplified. The test model has k factors labeled by j with $0 \leq j < k$. Each factor has v_j values represented as f_{hj} for $0 \leq h < v_j$. Elements a_{ij} comprise the N-by-k test array of strength t. Each a_{ij} takes one of the values f_{hj} $(0 \leq i < N)$.

6.2.1 All the Same Factor Value

When $v_j = 1$ for some j, then $a_{ij} = f_{0j}$ for all i, that is, all values are the same. Moreover, it does not matter whether the factor is included in the test case generation. When a covering test design has a single-value factor appended, it still covers: All the allowed $(t - 1)$-tuples from the original design are associated with the new value, so the new array covers with strength t. Thus, inclusion of such a factor is only a matter of convenience or clarity. In the design-your-burger test cases (Table 6.1), the submit button input is not included explicitly. It is required, however, to trigger the action of the form in every test case. Since the submit button input is a single-value factor, it can be added to every test case without affecting the t-way coverage of the test array.

6.2.2 All Different Factor Values

It is not unusual for some test factors to require unique values. That is, $a_{ij} \neq a_{i'j}$ for all $i \neq i'$ and some j. Examples include the selection of login names and database keys. Typically, these factors are limited to strength–1 coverage because each value can be used only once. Specifying uniqueness constraints in a strength-t design will not change this fact. A simpler approach with the same outcome is to substitute a single-value placeholder for the all-different factor during test case generation. Then, the placeholder value can be replaced with N unique values in the generated test cases. This process is illustrated in the database table example in Table 6.3.

6.2.3 Functionally Dependent Factor Values

When the values f_{hj} of factor j are determined by the values of l other factors $f_{h'j'}$ $(j' \neq j)$, factor j is functionally dependent on those factors [48]. If the l *determinant factors* are used to generate the test cases, the *dependent factor* value f_{hj} is fixed according to their values. The simplification is to represent f_{hj} with a placeholder value to generate the test cases and then plug in the values determined by the l factors.

TABLE 6.3 Reservation Database Records with Unique Room ID

Location ID	Room ID	Room Type	Section	Smoking	Beds	Sleep Sofa	Near Gym/Spa	Near Busn Center	Wi-Fi	Phone Lines	Balcony/ Patio
Location_1	Room_1	Single	Floor_1	No	Single	No	No	No	No	1	No
Location_2	Room_2	Suite	Mezzanine	Yes	2_queen	Yes	Yes	Yes	Yes	2	Yes
Location_3	Room_3	Standard	Tower	No	1_king	No	No	Yes	Yes	1	Yes
Location_3	Room_4	Standard	Floor_1	Yes	1_king	No	Yes	No	No	1	No
Location_2	Room_5	Standard	Tower	No	2_queen	No	No	No	No	1	No
Location_1	Room_6	Standard	Mezzanine	No	1_king	No	Yes	No	No	1	Yes
Location_1	Room_7	Suite	Tower	Yes	2_queen	Yes	Yes	Yes	Yes	2	Yes
Location_3	Room_8	Suite	Tower	No	2_queen	Yes	Yes	Yes	Yes	2	Yes
Location_2	Room_9	Single	Floor_1	Yes	Single	No	No	No	No	1	No
Location_3	Room_10	Single	Tower	No	Single	No	No	No	No	1	No
Location_3	Room_11	Standard	Mezzanine	No	2_queen	No	Yes	No	Yes	1	No
Location_2	Room_12	Standard	Floor_1	No	1_king	No	Yes	Yes	No	1	Yes
Location_1	Room_13	Standard	Floor_1	Yes	2_queen	No	No	Yes	Yes	1	Yes
Location_1	Room_14	Standard	Tower	Yes	2_queen	No	Yes	Yes	Yes	1	No
Location_2	Room_15	Standard	Mezzanine	Yes	2_queen	No	No	Yes	Yes	1	Yes

Source: Courtesy of Testcover.com. www.testcover.com.

Note: These test values represent rooms to be added to a reservation database table for new motels at three locations. The pairwise design includes constraints so that the test records conform to business rules for new features and amenities. The Room ID column uniquely identifies each room in the table. The single value room_ID was used as a placeholder during test case generation. The unique values shown were substituted afterwards. All pairs allowed by the business rules are included, except that each Room ID value is paired with only one value of each of the other columns.

TABLE 6.4 Partition for Valid Calendar Dates Using Functionally Dependent Last Day

```
Calendar Example using functionally dependent last day
Month
Day
Year
# valid dates partition
+ all months; first, tenth and last days; all years
jan feb mar apr may jun jul aug sep oct nov dec
1 10 last_day(month,year)
2015 2016 2017
```

Source: Courtesy of Testcover.com. www.testcover.com.

Note: This partition differs from that of Table 6.2 because it uses the placeholder value last_day(month,year) rather than explicit last day values (28 29 30 31). The partition contains only one block instead of five, but it requires substitution of the last day values into the generated test cases instead.

Representing the dependent factor values with a placeholder can simplify the expression of constraints so that fewer blocks are required. In the calendar example of Table 6.2, five blocks were needed in a partition for valid dates because of the variation in lengths of months. However, the fact that the last day of the month is functionally dependent on the month and year suggests an alternate partition with only one block, as in Table 6.4. The actual last days must be substituted into the test cases after generation.

In the case when *all* of the values of factor j have the same functional dependence and when $l \leq t$, all allowed l-tuples of the determinant factors will be covered, so the dependent factor j will include all its allowed values. Moreover, each l-tuple is associated with all allowed $(t - l)$-tuples of the nondeterminant factors (as is the corresponding value of the dependent factor). Thus, after substitution, the subarray of the nondeterminant factors and the dependent factor cover with strength $t - l + 1$.

More generally, when only *some* of the values of factor j are functionally dependent, the same simplification can be used (as in Table 6.4). Also, when the values of a factor are determined by *different* functions, different placeholders representing the functions can be used. Examples of these simplifications will be given in Section 6.4.

6.2.4 Hybrid Factor Values

Two or more factors can always be represented as a single hybrid factor [219] whose values are all of the allowed products of the original factors' values (see also Section 3.3.2). Thus, factors j and j' with v_j and $v_{j'}$ values, respectively, can be represented by a new factor with $\leq v_j v_{j'}$ values. This simplification can reduce the number of blocks required. However, the usual tradeoff is an increased number of test cases.

For example, three factors A, B, and C with the allowed values 0, 1, 2, 3 have the constraint $A < B$. The allowed combinations of A, B, and C can be specified in three blocks which lead to 16 pairwise test cases. Alternatively, the hybrid factor AB with values 01, 02, 03, 12, 13, 23 paired with C in one block yields 24 test cases.

6.3 SEQUENCE UNIT REPLAY MODEL

The sequence unit replay model applies combinatorial design to test a sequence of state transitions with comparisons to expected results. Each partition corresponds to a state transition path such that equivalent behavior is expected at each transition. Combinations of parameters are selected to replay different runs. The partition's test cases all follow the same path and cover all allowed t-tuples. The state transition paths are chosen to meet test goals.

Each *sequence unit* consists of test inputs (e.g., from one or more test automation scripts), which are run as a contiguous unit during test execution. Sequence units may have parameters to specify which inputs are exercised, and with which values. Examples of sequence units for an online shopping cart user interface (Figure 6.2) are listed in Table 6.5.

The test factors are sequence unit replay slots, ordered chronologically, to exercise the system behavior of interest. Thus, earlier sequence units not only test the system; they also set up the system for later interactions.

Instant Shopping

Your cart contains:

Delete	Item Number	Item Description	Quantity	Price	Item Total
☐	itemA	descriptionA	1	14.95	14.95
☐	itemB	descriptionB	2	9.95	19.90
☐	itemC	descriptionC	1	5.95	5.95
					+
				Subtotal:	$ 40.80

< Shop   Update   Checkout >

FIGURE 6.2 Shopping cart containing three different items. Items can be checked for deletion, and new quantities can be entered. The update button initiates these changes; the shop button returns to the previous shopping page; the checkout button presents the user interface for payment and shipping. An empty cart does not provide buttons for update or checkout.

TABLE 6.5 Shopping Cart Sequence Units

Sequence Unit	Replay Action
`suShopA`	`Go to shopping page for itemA and add it to the shopping cart. Go to the cart.`
`suShopB`	`Go to shopping page for itemB and add it to the shopping cart. Go to the cart.`
`suShopC`	`Go to shopping page for itemC and add it to the shopping cart. Go to the cart.`
`suCart`	`Go to the cart.`
`suDelete(i)`	`Check (or uncheck) the cart delete box for the item in position` *i*.
`suNewQ(i,q)`	`Type` *q* `in the cart quantity box for the item in position` *i*.
`suUpdate`	`Click the cart update button.`
`suCheckout`	`Click the cart checkout button.`

Source: Courtesy of Testcover.com. www.testcover.com.

Note: Test automation sequence units and their parameters are the factor values in a sequence unit replay model. These sequence units are the basis of the shopping cart test design.

Generally, test inputs and result checks may use multiple interfaces, as long as the timing of interactions is coordinated to be repeatable. Table 6.6 shows the test factors and values for the shopping cart example in DPB notation. In this partition, the slots correspond to placing different items into the cart, possibly checking them for deletion or entering new quantities, and possibly updating the cart and proceeding to checkout. Each block is constrained not to have an empty cart before update and before checkout, because empty carts cannot be updated or proceed to checkout. Table 6.7 illustrates test cases generated for this shopping cart sequence unit replay design. All allowed pairs of the selected test factors are covered.

The success of a sequence unit replay design depends on each test case of a partition taking the SUT through a sequence of equivalent state transitions (one path, one behavior). This is a necessary condition to cover the allowed combinations of inputs. Otherwise some combinations may not be used as intended, as in the design-your-burger example. State transition paths may be selected in different ways to meet this requirement. If a state machine diagram is available, the state transition paths can be chosen systematically, using known state machine test methods [16].

Because the sequence unit replay model does not dictate the transition paths, it offers flexibility. It can be applied to selected SUT behaviors to support goals like the following.

TABLE 6.6 Shopping Cart Test Partition for Sequence Unit Replay Model

```
Shopping Cart Example - sequence unit replay model
Position0
Position1
Position2
Delete0
Delete1
Delete2
NewQ0
NewQ1
NewQ2
Update
Checkout
#
+ n=1, no suDelete(0), no suNewQ(0,0)
suShopA suShopB suShopC
n/a
n/a
n/a
n/a
n/a
n/a suNewQ(0,1) su NewQ(0,2) suNewQ(0,10)
n/a
n/a
n/a suUpdate
n/a suCheckout
```

```
+ n=2, no suDelete(1), no suNewQ(1,0)
suShopA suShopB
suShopC
n/a
n/a suDelete(0)
n/a
n/a
n/a suNewQ(0,0) suNewQ(0,1) suNewQ(0,2) suNewQ(0,10)
n/a suNewQ(1,1) suNewQ(1,2) suNewQ(1,10)
n/a
n/a suUpdate
n/a suCheckout
+ n=3, no suDelete(2), no suNewQ(2,0)
suShopA
suShopB
suShopC
n/a suDelete(0)
n/a suDelete(1)
n/a
n/a suNewQ(0,0) suNewQ(0,1) suNewQ(0,2) suNewQ(0,10)
n/a suNewQ(1,0) suNewQ(1,1) suNewQ(1,2) suNewQ(1,10)
n/a suNewQ(2,1) suNewQ(2,2) suNewQ(2,10)
n/a suUpdate
n/a suCheckout
```

Note: Test factors correspond to sequence unit replay slots. One to three different items are placed into the cart in the first three slots. Items are checked for deletion and have new quantities entered in the next six slots. The cart may be updated and checked out in the last two slots. The three blocks handle the cases of one, two, and three initial items, respectively. Each block is constrained not to have an empty cart before update and before checkout. A factor with the value n/a indicates that a sequence unit is not run in that slot.

TABLE 6.7 Test Cases for Shopping Cart Sequence Unit Replay Model

Position0	Position1	Position2	Delete0	Delete1	Delete2	NewQ0	NewQ1	NewQ2	Update	Checkout
suShopB	n/a	n/a	n/a	n/a	n/a	suNewQ(0,10)	n/a	n/a	suUpdate	suCheckout
suShopA	suShopB	suShopC	suDelete(0)	suDelete(1)	n/a	suNewQ(0,1)	suNewQ(1,10)	suNewQ(2,1)	n/a	n/a
suShopA	suShopC	n/a	suDelete(0)	n/a	n/a	suNewQ(0,0)	suNewQ(1,1)	n/a	n/a	suCheckout
suShopA	suShopB	suShopC	n/a	suDelete(1)	n/a	n/a	suNewQ(1,2)	suNewQ(2,10)	suUpdate	n/a
suShopA	suShopB	suShopC	suDelete(0)	n/a	n/a	suNewQ(0,0)	suNewQ(1,0)	suNewQ(2,2)	suUpdate	n/a
suShopB	suShopC	n/a	n/a	n/a	n/a	suNewQ(0,2)	suNewQ(1,2)	n/a	n/a	n/a
suShopC	n/a	n/a	n/a	n/a	n/a	suNewQ(0,1)	n/a	n/a	suUpdate	n/a
suShopA	suShopB	suShopC	suDelete(0)	suDelete(1)	n/a	suNewQ(0,10)	n/a	suNewQ(2,10)	suUpdate	suCheckout
suShopA	suShopC	n/a	suDelete(0)	n/a	n/a	n/a	suNewQ(1,10)	n/a	suUpdate	suCheckout
suShopA	suShopB	suShopC	suDelete(0)	suDelete(1)	n/a	suNewQ(0,2)	suNewQ(1,2)	suNewQ(2,1)	suUpdate	suCheckout

Note: The first 10 of the 35 shopping cart test cases are shown. All test cases start with an empty cart. In the first test case itemB is placed in cart position0; the quantity of this item is set to 10; the cart is updated and then sent to checkout.

- Test the delivered functions of an incomplete system.
- Test a system including externally provided components not specified in detail.
- Test a system for which operational specifications are not complete or up to date.
- Focus testing on behaviors of particular importance.

To apply the sequence unit replay model in these situations, test designers may need to construct or modify a state machine diagram so that the test design conforms to applicable requirements.

The choice of test model needs to reflect appropriate tradeoffs among project goals, resources, and risks. The flexibility of transition path selection in this model comes with a risk of omission of some transitions. The test models presented in the rest of this chapter include all transitions to avoid this risk. For comparison, the next section presents a combinatorial test design for the same online shopping cart.

6.4 SINGLE REGION STATE MODELS

This section and the next one describe combinatorial test models based on UML state machine diagrams. UML state machines are an extension of traditional finite state machines: They can have hierarchically nested states and actions associated with the entry and exit of the states. A UML state machine also can describe multiple regions operating concurrently. Each region of a UML state machine has its own states, and the overall operation reflects all the regions interacting with each other and external stimuli. The test models described here depend not only on the diagrammed state, but also on associated program variables, which may include inputs from set-up states, database values, and so on. Together the state and program variables define the *extended state*, which must be set correctly for repeatable test results.

Behaviors of a state machine region are defined by its state transitions, so testing the SUT models involves testing transitions. This consists of setting up the source (from) state, triggering the transition, and observing what happens. The expected result is the action of the transition to the target (to) state indicated by the state machine diagram. Test factors include the source state, associated program variables, and the event expected to trigger the transition. Execution of each test case involves applying a sequence of set-up inputs to the SUT so that it is in the specified source state with program variables set to the test case values. With the extended state set, the trigger is applied and results observed.

The online shopping cart state diagram is shown in Figure 6.3. Tables 6.8 through 6.10 describe actions in the diagram using pseudocode. The shopping cart has two lowest-level *leaf* states—emptyCart and nonempty-Cart; transitions to/from these states, as well as the shopping and checkout states, are to be tested. There are eight possible transitions among the four states, as shown in the left column of Table 6.11. The *transition model* tests each transition in its own partition. In this model, the partition is associated only with the source and target states of its transition. (The source state is a single-value factor; the expected target state is determined by the event factor.) Consequently, all the *t*-tuples in the partition also are associated with the source and target states.

A second model—the *target state model*—follows when the eight transitions are grouped into four partitions according to their target states (right column of Table 6.11). In the target state model, all *t*-tuples in each partition are associated with the corresponding target state. Each source state is covered with strength *t*, but it might not be associated with *all* the

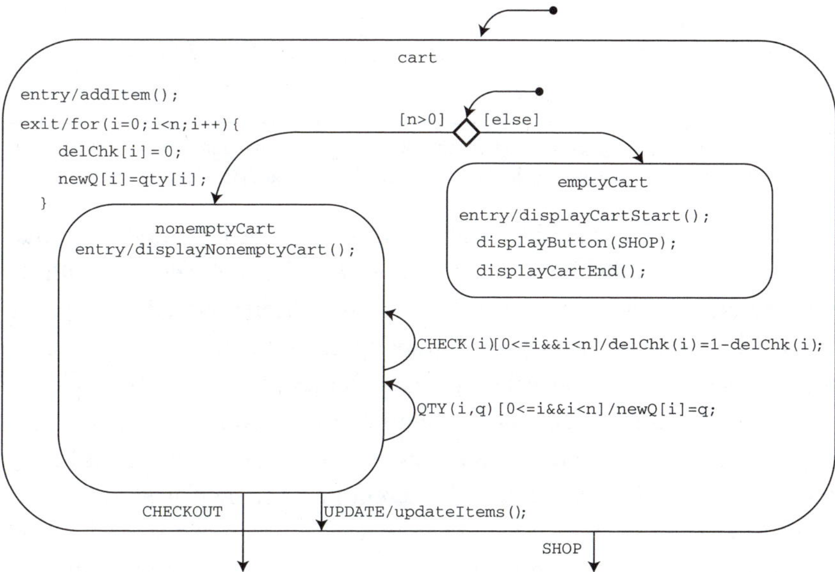

FIGURE 6.3 Shopping cart state diagram. Transitions among the leaf states (nonemptyCart and emptyCart) and the shopping and checkout states are to be tested. Pseudocode for the actions addItem(), displayNonemptyCart(), and updateItems() is given in Tables 6.8 through 6.10.

TABLE 6.8 Pseudocode for addItem() Action on Entry to the cart State in Figure 6.3

```
addItem() {     /* cart entry */
     if(!isset(n)) n=0;
     if(isset(newItem)) {
          for(i=0; i<n; i++) {
                if(newItem==item[i]) {
                     /* increment quantity of item in cart */
                     qty[i]++;
                     newQ[i]=qty[i];
                     itemTotal[i]=qty[i]*price[i];
                     break;
                }
                if(i==n-1) {
                     /* item not in cart - look in catalog */
                     select ctlgItem, ctlgDesc, ctlgPrice from ctlg
                         where ctlgItem=newItem;
                     if(isset(ctlgItem)) {
                          delChk[n]=0;
                          item[n]=ctlgItem;
                          desc[n]=ctlgDesc;
                          qty[n]=1;
                          newQ[n]=1;
                          price[n]=ctlgPrice;
                          itemTotal[n]=ctlgPrice;
                          n++;
                          unset(ctlgItem);
                     }
                }
          }
          unset(newItem);
     }
}
```

Note: When a new item is already in the shopping cart, its quantity is incremented. Otherwise, the new item is looked up in the catalog database.

t-tuples in the partition. The transition model can be more thorough than the target state model because all allowed *t*-tuples are exercised with the source state, as well as the target state. The transition model designs generally yield more test cases also: In the shopping cart example, the transition model design has 187 pairwise test cases; the target state design has 172.

Test factors and values for the shopping cart example, are shown in Table 6.12. The same factors and values are used in both models. As before, the test design allows up to three different items to be placed in the cart. Most of the test factors are program variables from the state diagram. (These may or may not correspond to variables in the implementation.) The state and event factors are respectively the source state and the trigger for the transition to the target state.

There are 16 blocks for transitions to the nonemptyCart state. In the target state model, they are grouped into one partition as follows. The first four

TABLE 6.9 Pseudocode for displayNonemptyCart() Action on Entry to the nonemptyCart State in Figure 6.3

```
displayNonemptyCart() {/* nonemptyCart entry */
      displayCartStart();
      subtotal=0;
      for(i=0; i<n; i++) {
            displayItem(delChk[i], item[i], desc[i], newQ[i],
                  price[i], itemTotal[i]);
            subtotal+=itemTotal[i];
      }
      displaySubtotal(subtotal);
      displayButton(SHOP);
      displayButton(UPDATE);
      displayButton(CHECKOUT);
      displayCartEnd();
}
```

Note: Each item in the cart is displayed; the subtotal and the SHOP, UPDATE, and CHECKOUT buttons are displayed, as shown in Figure 6.2.

TABLE 6.10 Pseudocode for updateItems() Action on the UPDATE Transition from the nonemptyCart State to Its Containing State cart in Figure 6.3

```
updateItems() {      /* nonemptyCart UPDATE */
      newI=0;
      for(i=0; i<n; i++) {
            if(delChk[i]==1||newQ[i]==0) continue;
                  /* skip removed items */
            delChk[newI]=0;
            item[newI]=item[i];
            desc[newI]=desc[i];
            if(newQ[i]>0) {
                  /* update valid quantities */
                  qty[newI]=newQ[i];
                  newQ[newI}=newQ[i];
            }
            else {
                  /* ignore invalid quantities */
                  qty[newI]=qty[i];
                  newQ[newI]=qty[i];
            }
            price[newI]=price[i];
            itemTotal[newI]=qty[newI]*price[newI];
            newI++;
      }
      n=newI;
}
```

Note: Removed items are skipped; items with valid new quantities are updated; item totals are recomputed.

TABLE 6.11 Partitions for the Shopping Cart State Diagram of Figure 6.3

Transition Model	Target State Model
`# shopping to emptyCart` `# nonemptyCart to emptyCart`	`# all states to emptyCart`

`# shopping to nonemptyCart`
`# nonemptyCart to nonemptyCart`
`# checkout to nonemptyCart`
	`# all states to nonemptyCart`

`# emptyCart to shopping` `# nonemptyCart to shopping`	`# all states to shopping`
`# nonemptyCart to checkout`	`# all states to checkout`

Note: In the transition model, each allowed transition has its own partition. In the target state model, all the transitions to each target state are included in one partition. The same factors, values, and blocks are used in both models. For example, the blocks comprising the three transitions to the nonemptyCart state in the transition model are grouped into one partition for the target state model.

blocks are from the shopping state when there are 0–3 different items already in the cart. (These blocks comprise one partition in the transition model.) The next nine blocks are from and to the nonemptyCart state. The CHECK, QTY, and UPDATE events each use three blocks for when there are 1–3 items already in the cart. The last three blocks are from the checkout state to the nonemptyCart. Table 6.13 shows the first 10 of 74 test cases for the nonemptyCart partition in the target state model. These test cases can be set up and executed using automation scripts like those of the sequence units in Table 6.5. The design details for both models, including the DPB notation for their partitions, are available in the Testcover.com tutorial [197].

The transitions to the nonemptyCart state provide a few examples of constraint simplification with functionally dependent values. First, the exit action of the cart state (Figure 6.3) resets newQ[*i*], the new quantity shown in cart position *i* (which may have been be changed), back to qty[*i*], the current quantity of that item. Thus, in normal operation, the value of newQ[0] is constrained to be the same as qty[0] in transitions back to the nonemptyCart state from the shopping state. Similar constraints occur for newQ[1] and newQ[2], as well as for transitions back from the checkout state. Here blocks are not split to make the values of newQ[0] match those of qty[0] for these transitions: Test cases are generated with qty[0] as the value for newQ[0], and the corresponding value is substituted afterwards. Similar substitutions are made for newQ[1] and newQ[2]. Different examples of functionally dependent values occur in the transition from and to the nonemptyCart state. Here, there are three events, CHECK(*i*), QTY(*i*,*q*), and

TABLE 6.12 Test Factors and Values for the Shopping Cart State Diagram of Figure 6.3

Test Factor	Test Factor Values	Indication
newItem	unset itemA itemB itemC	Item to place in cart
n	unset 0 1 2 3	Number of items in cart
delChk[0]	n/a 0 1	Delete box checked in cart position 0
item[0]	n/a itemA itemB itemC	Item in cart position 0
qty[0]	n/a 1 2 10	Quantity of item in cart position 0
newQ[0]	n/a 0 1 2 10 qty[0]	New quantity shown in cart position 0
delChk[1]	n/a 0 1	Delete box checked in cart position 1
item[1]	n/a itemB itemC	Item in cart position 1
qty[1]	n/a 1 2 10	Quantity of item in cart position 1
newQ[1]	n/a 0 1 2 10 qty[1]	New quantity shown in cart position 1
delChk[2]	n/a 0 1	Delete box checked in cart position 2
item[2]	n/a itemC	Item in cart position 2
qty[2]	n/a 1 2 10	Quantity of item in cart position 2
newQ[2]	n/a 0 1 2 10 qty[2]	New quantity shown in cart position 2
i	n/a 0 1 2	Cart position for event
q	n/a 0 1 2 10	Quantity for event
state	shopping emptyCart nonemptyCart checkout	Source state
event	SHOP CART CHECK(i) [0<=i&&i<n] QTY(i,q) [0<=i&&i<n] UPDATE CHECKOUT	Trigger to target state

Note: The first 16 factors are program variables controlling the set-up and test of the cart transition. The last two factors are the source state and the transition event. The expected result is the transition to the target state. In this example up to three different items are placed into the cart. A factor with the value n/a indicates that the factor is not applicable to the transition test case.

UPDATE. (*q* is the candidate new quantity for cart position *i*.) The action of CHECK(*i*) is determined by *i*; the action of QTY(*i*,*q*) is determined by both *i* and *q*; but UPDATE does not depend on either.

Both the transition model and the target state model follow the state machine diagram systematically. Their use ensures coverage of all state transitions defined by the state machine. In this example, the sequence

TABLE 6.13 Test Cases for Shopping Cart Target State Model

newItem	*n*	delChk[0]	item[0]	qty[0]	newQ[0]	delChk[1]	item[1]	qty[1]	newQ[1]	delChk[2]	item[2]	qty[2]	newQ[2]	*i*	*q*	state	event
itemA	unset	n/a	n/a	n/a	qty[0]	n/a	n/a	n/a	qty[1]	n/a	n/a	n/a	qty[2]	n/a	n/a	shopping	CART
unset	3	0	itemA	1	0	0	itemB	1	0	0	itemC	1	0	0	n/a	nonemptyCart	CHECK(i)[0<=i&&i<n]
unset	2	1	itemB	2	0	1	itemC	2	1	n/a	n/a	n/a	n/a	1	1	nonemptyCart	QTY(i,q)[0<=i&&i<n]
unset	3	1	itemA	10	2	1	itemB	10	2	1	itemC	10	2	2	2	nonemptyCart	QTY(i,q)[0<=i&&i<n]
unset	1	0	itemC	10	10	n/a	n/a	n/a	n/a	n/a	n/a	n/a	n/a	n/a	n/a	nonemptyCart	UPDATE
unset	3	0	itemA	2	qty[0]	0	itemB	2	qty[1]	0	itemC	2	qty[2]	n/a	n/a	checkout	CART
itemC	2	0	itemB	1	qty[0]	0	itemC	10	qty[1]	n/a	n/a	n/a	qty[2]	n/a	n/a	shopping	CART
unset	3	1	itemA	2	10	1	itemB	2	10	1	itemC	2	10	1	10	nonemptyCart	QTY(i,q)[0<=i&&i<n]
unset	3	1	itemA	1	1	1	itemB	1	1	1	itemC	1	1	0	1	nonemptyCart	QTY(i,q)[0<=i&&i<n]
unset	1	0	itemB	1	2	n/a	n/a	n/a	n/a	n/a	n/a	n/a	n/a	0	0	nonemptyCart	QTY(i,q)[0<=i&&i<n]

Note: There are 74 test cases in the nonemptyCart partition; the first 10 are shown here. In the first test case, itemA is placed in the cart, from the shopping state. The second test case is set up with three items in the cart with their new quantities set to 0; the delete box is checked in cart position 0 in this transition from and to the nonemptyCart state. In the transitions from shopping or checkout, newQ[0] is determined by qty[0] because newQ[0] = qty[0]. Similarly, newQ[1] and newQ[2] are determined by qty[1] and qty[2], respectively. In the transitions from nonemptyCart, CHECK(*i*) is determined by *i*, and QTY(*i*,*q*) is determined by *i* and *q*.

unit replay design tests much of the shopping cart behavior, but it has no transition to an empty cart, no return from checkout, and so on. The transition model and the target state model do cover these transitions, and require more test cases to do so.

6.5 MULTIPLE REGION STATE MODELS

When a state machine diagram has more than one region, a useful starting point is to test each region as if isolated from the rest of the system. These *stand-alone* designs can use the transition model and the target state model described in the previous section. When the behaviors of a region explicitly depend on another region (e.g., its state or a message), corresponding factors can be included. However, these stand-alone designs do not address the concurrent operation of the other regions generally.

This section introduces three models for *integrated* designs, which test each region's behaviors and its interactions with the other regions. The models are integrated in that states of all regions are included as factors in testing each region. The integrated transition model and the integrated target state model use their respective partitioning as before, but the integrated test designs control the states of the other regions as well. The third model, the *integrated propagation model*, focuses on the observable effects of a region's transitions on *other* regions, using fewer test cases. It uses the same partitioning and factors as the integrated target state model. However, the only combinations of values selected are for interactions with visible results in other regions. State machine transitions that have no observable effect in other regions are excluded.

Generally, integrated model test designs are based on corresponding stand-alone designs for each region, with current states of other regions as new factors. A typical first approach is to allow all combinations of states. However, some combinations may not be feasible if the system design prohibits them or makes their set-up too difficult. (The test cases specify all the regions' states when the event occurs. The risks due to test omissions should be considered if the design is constrained to exclude combinations of states.) After the state combinations are chosen, program variable values are selected for each block, consistent with all the regions' current states and the corresponding events. Additional detail on the design procedure is given in the Testcover.com tutorial [197].

The test designs described here are for a thermostat (Figure 6.4), simplified for the purpose of this example. The thermostat state machine diagram (Figure 6.5) has three orthogonal (concurrent) regions: The tempSet region

FIGURE 6.4 Thermostat panel. The thermostat controls heating, cooling, and fan functions to maintain room temperature according to user settings. Normally, the thermostat displays the current room temperature. The SET button is used to select a state to display and set the temperature for heating or cooling. The temperature setting is changed by pressing the UP or DOWN button. The temperature control mode is selected with a sliding switch. In the OFF mode, there is no heating or cooling. In the HEAT mode, the thermostat turns the furnace on when the room temperature is less than the heat setting, and off when the room temperature exceeds the heat setting by a small increment. In the COOL mode, the air conditioner is operated similarly. Control of the air circulation fan is set with a second sliding switch. In the AUTO position, the fan is on whenever the furnace or air conditioner is on. In the ON position, the fan circulates air continually.

allows the user to set the heating and cooling temperatures. Transitions among the leaf states (idleWait, heatKeyWait, and coolKeyWait) are to be tested. The tempControl region controls the heating and cooling according to the operating mode selected by the user. Transitions among the leaf states (heatOffWait, heatOnWait, coolOffWait, coolOnWait, and tempControlOff) are to be tested. The fanControl region controls the operation of the fan circulating the air. Transitions among the leaf states (autoOffWait, autoOnWait, and fanOn) are to be tested.

The numbers of test cases generated in the different thermostat example designs are shown in Table 6.14. In particular, the tempControl integrated target state design has 66 test cases in five partitions. There are 16 allowed state transitions grouped into five partitions according to their target states: heatOffWait, heatOnWait, coolOffWait, coolOnWait, and

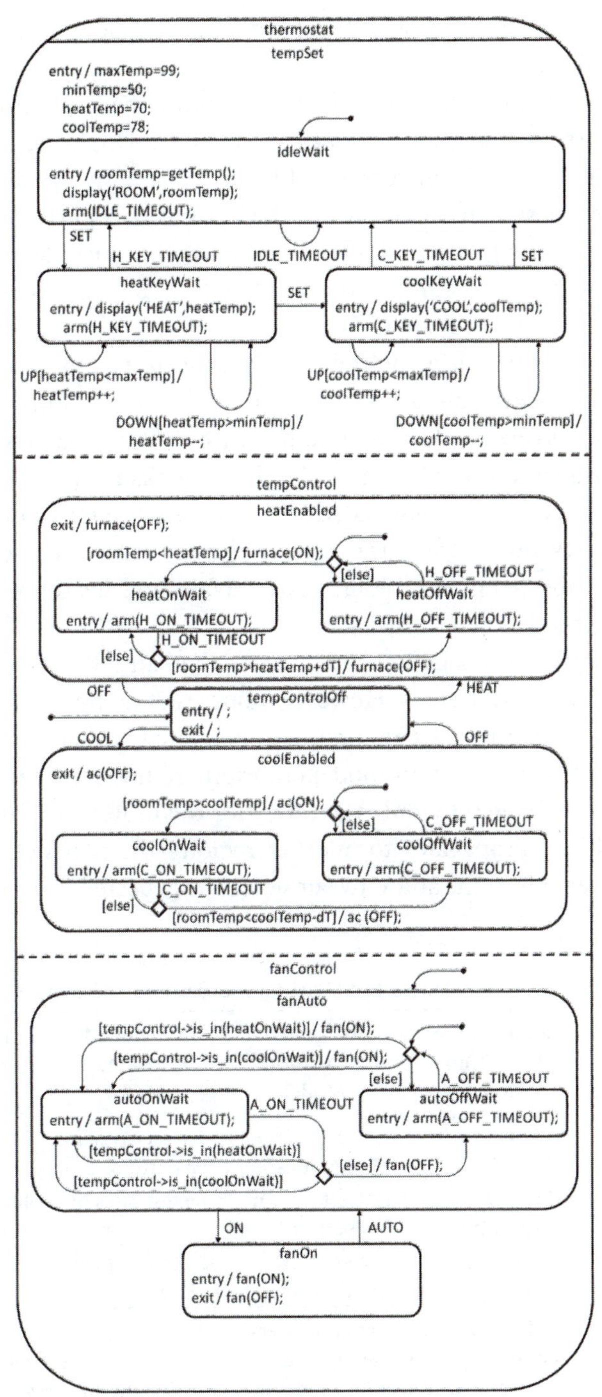

FIGURE 6.5 Thermostat state diagram with three regions.

tempControlOff. In this design, there are seven test factors with values as given in Table 6.15.

The first partition includes all transitions to the heatOffWait state. There are three possible source states: tempControlOff, heatOffWait, and heatOnWait. The partition consists of three blocks corresponding to the source states, as shown in Table 6.16. The heatOffWait partition yields 12 pairwise test cases given in Table 6.17. This design includes all allowed pairs of current states leading to the target state. Each event value is paired with each of the allowed current states.

The integrated transition design uses the same test factors and values as the integrated target state design for the tempControl region (Table 6.15). Moreover, the same blocks are used, but with different partitioning. The integrated transition design places the three blocks from Table 6.16 into separate partitions, each associated with a different tempControl source state. In this test design there are respectively 9, 6, and 6 test cases for transitions from the tempControlOff, heatOffWait, and heatOnWait source states to the heatOffWait target state.

The integrated propagation design uses the same test factors as the other two integrated designs. However, values that do not lead to observable effects in other regions are omitted. Integrated transition blocks that do not meet the propagation model criterion are not used either. Of the three blocks in Table 6.16, only the third (for the heatOnWait to heatOffWait transition) propagates to another region—the fanControl region, and only when it is in the autoOnWait state. The expected transition in the

TABLE 6.14 Numbers of Pairwise Test Cases for the Thermostat Example Test Designs

Orthogonal Region	Stand-Alone Transition Model	Integrated Transition Model	Integrated Target State Model	Integrated Propagation Model
tempSet	65 test cases in 7 partitions	105 test cases in 7 partitions	76 test cases in 3 partitions	34 test cases in 3 partitions
tempControl	64 test cases in 16 partitions	120 test cases in 16 partitions	66 test cases in 5 partitions	40 test cases in 5 partitions
fanControl	15 test cases in 8 partitions	96 test cases in 8 partitions	52 test cases in 3 partitions	

Source: Courtesy of Testcover.com. www.testcover.com

Note: In each of the three regions, the same test factor values (i.e., the same blocks) are used for the integrated transition model design as for the integrated target state model design. Differences in numbers of test cases are due to differences between the models, not different choices of test values.

TABLE 6.15 Test Factors and Values for the Thermostat tempControl Region Integrated Target State Design

Test Factor	Test Factor Values
roomTemp	49 70 78 100
heatTemp	50 69 79 99
coolTemp	50 69 79 99
tempSet state	idleWait heatKeyWait coolKeyWait
tempControl state	heatOffWait heatOnWait coolOffWait coolOnWait tempControlOff
fanControl state	autoOffWait autoOnWait fanOn
event	HEAT[roomTemp>=heatTemp] HEAT[roomTemp<heatTemp] H_OFF_TIMEOUT[roomTemp>=heatTemp] H_OFF_TIMEOUT[roomTemp<heatTemp] H_ON_TIMEOUT[roomTemp>heatTemp+dT] H_ON_TIMEOUT[roomTemp<=heatTemp+dT] COOL[roomTemp<=coolTemp] COOL[roomTemp>coolTemp] C_OFF_TIMEOUT[roomTemp<=coolTemp] C_OFF_TIMEOUT[roomTemp>coolTemp] C_ON_TIMEOUT[roomTemp<coolTemp-dT] C_ON_TIMEOUT[roomTemp>=coolTemp-dT] OFF

Note: The first three factors are program variables for the room temperature and the heating and cooling settings. The next three factors are the current states of the three regions when the tempControl trigger in the last factor occurs. The expected result is the transition to the target state in the tempControl region of the state machine diagram.

fanControl region is from autoOnWait to autoOffWait. This more restrictive model leads to six test cases for transitions to the heatOffWait target state.

The example of transitions to the heatOffWait state in the tempControl region illustrates the use of different test models for different purposes. Integrated designs based on state machines with multiple regions support testing while the states of all regions are controlled. The transition model, target state model, and propagation model offer a choice of designs depending on test goals.

6.6 CHAPTER SUMMARY

Controlling the SUT state is essential to knowing what is being tested and how thoroughly. Attention to the state of the system can enable the tester

TABLE 6.16 Thermostat heatOffWait Partition in tempControl Integrated Target State Design

```
Thermostat Example-tempControl integrated target state design
roomTemp
heatTemp
coolTemp
tempSet state
tempControl state
fanControl state
event
#TC:h tempControl states to heatOffWait
+tempControlOff to heatOffWait
70 100
50 69
50 69
idleWait heatKeyWait coolKeyWait
tempControlOff
autoOffWait autoOnWait fanOn
HEAT[roomTemp>=heatTemp]
+heatOffWait to heatOffWait
70 100
50 69
50 69
idleWait heatKeyWait coolKeyWait
heatOffWait
autoOffWait fanOn
H_OFF_TIMEOUT[roomTemp>=heatTemp]
+heatOnWait to heatOffWait
70 100
50 69
50 69
idleWait heatKeyWait coolKeyWait
heatOnWait
autoOnWait fanOn
H_ON_TIMEOUT[roomTemp>heatTemp+dT]
```

Note: The three blocks specify the allowed tempControl transitions from tempControlOff, heatOffWait, and heatOnWait, respectively. Any of the current tempSet states are allowed for these transitions. In the heatOffWait state (second block), the fanControl state is constrained not to be autoOnWait. Similarly, in the heatOnWait state (third block), the fanControl state is constrained not to be autoOffWait.

to design a more complete plan for testing more states and to apply the test inputs more accurately with appropriate partitioning. Test factor partitions can be defined to set up inputs that can reach each source state, to be consistent with the trigger and guard conditions, and to verify boundary values.

TABLE 6.17 Test Cases for Thermostat HeatOffWait Integrated Target State Design

roomTemp	heatTemp	coolTemp	tempSet State	tempControl State	fanControl State	event
70	69	50	coolKeyWait	tempControlOff	fanOn	HEAT[roomTemp>=heatTemp]
100	69	69	heatKeyWait	heatOffWait	autoOffWait	H_OFF_TIMEOUT[roomTemp>=heatTemp]
70	50	69	idleWait	heatOnWait	autoOnWait	H_ON_TIMEOUT[roomTemp>heatTemp+dT]
100	50	50	heatKeyWait	tempControlOff	autoOnWait	HEAT[roomTemp>=heatTemp]
100	69	50	idleWait	heatOnWait	fanOn	H_ON_TIMEOUT[roomTemp>heatTemp+dT]
70	50	50	heatKeyWait	heatOffWait	fanOn	H_OFF_TIMEOUT[roomTemp>=heatTemp]
100	69	69	coolKeyWait	tempControlOff	autoOffWait	HEAT[roomTemp>=heatTemp]
100	50	50	idleWait	tempControlOff	autoOffWait	HEAT[roomTemp>=heatTemp]
100	50	50	coolKeyWait	heatOnWait	autoOnWait	H_ON_TIMEOUT[roomTemp>heatTemp+dT]
70	69	69	coolKeyWait	heatOffWait	fanOn	H_OFF_TIMEOUT[roomTemp>=heatTemp]
70	50	69	idleWait	heatOffWait	autoOffWait	H_OFF_TIMEOUT[roomTemp>=heatTemp]
100	69	69	heatKeyWait	heatOnWait	autoOnWait	H_ON_TIMEOUT[roomTemp>heatTemp+dT]

Note: These 12 cases test the transitions from all allowed tempControl states to the heatOffWait state. All allowed pairs of current states leading to the heatOffWait state are included. Each event value is paired with each of the allowed current states. The first test case is set up by setting the temperature sensor simulator to 70, heatTemp to 69, and coolTemp to 50. The mode switch is set to OFF, and the fan switch is set to ON. After the tempSet, tempControl, and fanControl, states are idleWait, tempControlOff, and fanOn, the SET button is pressed twice to take the tempSet state to coolKeyWait. The mode switch is moved from OFF to HEAT before the coolKeyWait timeout occurs.

Combinatorial test models range from informal handling of SUT state, without requiring a complete specification, to test models based on transitions of state machines with one or more regions. The different models offer a range of tradeoffs for different development methods and goals.

Describing system operation in terms of state machines can help meet the challenge of incorporating partitions with constraints into combinatorial test designs: The state machine clarifies system behaviors and helps to avoid test design errors and omissions. Applying combinatorial testing to state machine regions is a systematic way to improve test designs, suggesting future benefits from its integration with tools for modeling, test automation, and test management.

REVIEW

Q: Why should invalid inputs for exception handling be partitioned separately from valid inputs?

A: The different inputs trigger different system behaviors and lead to different expected results. When inputs for different behaviors are mixed together, it is unclear what has been tested.

Q: Month, day, and year factors are included in two partitions, one for valid calendar dates and one for off-by-one boundary checking for months that are "too long." Which partition should contain February 29th?

A: Both partitions—with a leap year for the valid date and with a common year for the invalid date.

Q: An N-by-k covering array of strength t is constructed. Its factors are labeled 0 to $k-1$. Two new factors k and $k+1$ are appended to the array. The values of each of the new factors are determined by those of factor 0. What is the strength of the subarray of factors 1 to k? What is the strength of the subarray of factors 1 to $k-1$ and $k+1$?

A: Strength t; strength t.

Q: Four test factors with two values each describe the presence or absence of four inputs. The inputs are constrained such that at most one of the inputs is allowed. How many blocks are needed to describe the constraint with these factors? If the four factors are combined into a hybrid factor, how many values will it need to describe the allowed combinations in one block?

A: If exactly one input is required: four blocks (one for each input); four values. If the absence of all inputs is allowed: four blocks (one containing the no-input combination *and* an input); five values.

Q: Which test model applies combinatorial design to test inputs with informal management of the state of the SUT?

A: The sequence unit replay model.

Q: When the transitions to a target state have only one source state, what is the difference between the transition model and the target state model?

A: No difference.

Q: How many current state factors are there in an integrated test model? How many event factors?

A: One state factor for each region; one event factor to change the state of the region under test.

CHAPTER 7

Measuring Combinatorial Coverage

SINCE IT IS NEARLY always impossible to test all possible combinations, combinatorial testing is a reasonable alternative. But it is not always practical to redesign an organization's testing procedures to use tests based on covering arrays. Testing procedures often develop over time, and employees have extensive experience with a particular approach. Units of the organization may even be structured around established, documented test procedures. This is particularly true in organizations that must test according to contractual requirements, or regulatory standards. And because much software assurance involves testing applications that have been modified to meet new specifications, an extensive library of legacy tests will exist. The organization can save time and money by reusing existing tests, which generally have not been developed as covering arrays.

How can combinatorial methods be useful in such an environment? Short of creating new test suites from scratch, one of the best approaches to obtaining the advantages of methods described in this book is to measure the combinatorial coverage of existing tests, then supplement as needed. As introduced in Chapter 1, these methods help take advantage of the interaction rule, by testing combinations up to a suitable level. Depending on the budget and criticality of the software, two-way through five-way or six-way testing may be appropriate. Building covering arrays for some specified level of t is one way to provide t-way coverage. However, many large test suites naturally cover a high percentage of t-way combinations.

If our existing test suite covers almost all three-way combinations, for example, then it may be sufficient for the level of assurance that we need. Determining the level of input or configuration state space coverage can also help in understanding the degree of risk that remains after testing. If 90%–100% of the relevant state space has been covered, then presumably the risk is small, but if coverage is much smaller, then the risk may be substantial. This chapter describes some measures of combinatorial coverage that can be helpful in evaluating the degree of t-way coverage of any test suite, even when it was not constructed by combinatorial methods.

7.1 SOFTWARE TEST COVERAGE

Test coverage is one of the most important topics in software assurance. Users would like some quantitative measure to judge the risk in using a product. For a given test set, what can we say about the combinatorial coverage it provides? With physical products, such as light bulbs or motors, reliability engineers can provide a probability of failure within a particular time frame. This is possible because the failures in physical products are typically the result of natural processes, such as metal fatigue.

> Commonly used coverage measures do not apply well to combinatorial testing.

With software, the situation is more complex, and many different approaches have been devised for determining software test coverage.

With millions of lines of code, or only with a few thousand, the number of paths through a program is so large that it is impossible to test all paths. For each *if* statement, there are two possible branches, so a sequence of n *if* statements will result in 2^n possible paths. Thus, even a small program with only 270 *if* statements in an execution trace may have more than 10^{80} possible paths. With loops (*while* statements), the number of possible paths is literally infinite. Thus, a variety of measures have been developed to gauge the degree of test coverage. The following are some of the better-known coverage metrics:

- *Statement coverage*: This is the simplest of coverage criteria—the percentage of statements exercised by the test set. While it may seem at first that 100% statement coverage should provide good confidence in the program, in practice, statement coverage is a relatively weak criterion.

- *Decision or branch coverage:* The percentage of branches that have been evaluated to both *true* and *false* by the test set.

- *Condition coverage:* The percentage of conditions within decision expressions that have been evaluated to both true and false by the test set. Note that 100% condition coverage does not guarantee 100% decision coverage. For example, "`if (A || B) {do something} else {do something else}`" is tested with [0 1], [1 0], then A and B will both have been evaluated to 0 and 1, but the *else* branch will not be taken because neither test leaves both A and B false.

- *Modified condition decision coverage (MCDC):* This is a strong coverage criterion that is required by the U.S. Federal Aviation Administration for Level A (catastrophic failure consequence) software, that is, software whose failure could lead to complete loss of life. It requires that every condition in a decision in the program has taken on all possible outcomes at least once, and each condition has been shown to independently affect the decision outcome, and that each entry and exit point has been invoked at least once.

7.2 COMBINATORIAL COVERAGE

Note that the coverage measures above depend on access to program source code. Combinatorial testing, in contrast, is a black box technique. Inputs are specified and expected results determined from some form of specification. The program is then treated as simply a processor that accepts inputs and produces outputs, with no knowledge expected of its inner workings.

Even in the absence of knowledge about a program's inner structure, we can apply combinatorial methods to produce precise and useful measures. In this case, we measure the state space of inputs. Suppose we have a program that accepts two inputs, x and y, with 10 values each. Then, the input state space consists of the $10^2 = 100$ pairs of x and y values, which can be pictured as a checkerboard square of 10 rows by 10 columns. With three inputs, x, y, and z, we would have a cube with $10^3 = 1000$ points in its input state space. Extending the example to n inputs, we would have n dimensions with 10^n points. Exhaustive testing would require inputs of all 10^n combinations, but combinatorial testing could be used to reduce the size of the test set.

How should state space coverage be measured? Looking closely at the nature of combinatorial testing leads to several measures that are useful. We begin by introducing what will be called a *variable-value configuration*.

Definition

For a set of t variables, a variable-value configuration is a set of t valid values, one for each of the variables.

Example

Given four binary variables, a, b, c, and d, $a=0$, $c=1$, $d=0$ is a variable-value configuration, and $a=1$, $c=1$, $d=0$ is a different variable-value configuration for the same three variables a, c, and d. Choosing a different three-way combination of the variables, $b=0$, $c=1$, $d=0$ and $b=1$, $c=1$, $d=0$ are two different variable-value configuration for variables, b, c, and d.

7.2.1 Simple *t*-Way Combination Coverage

Of the total number of t-way combinations for a given collection of variables, what percentage will be covered by the test set? If the test set is a covering array, then coverage is 100%, by definition, but many test sets not based on covering arrays may still provide significant t-way coverage. If the test set is large, but not designed as a covering array, it is very possible that it provides two-way coverage or better. For example, random input generation may have been used to produce the tests, and good branch or condition coverage may have been achieved. In addition to the structural coverage figure, for software assurance, it would be helpful to know what percentage of two-way, three-way, and so on coverage has been obtained.

Definition

For a given test set for n variables, simple t-way combination coverage is the proportion of t-way combinations of n variables for which all variable-value configurations are fully covered.

TABLE 7.1 An Example Test Array for a System with Four Binary Variable

a	*b*	*c*	*d*
0	0	0	0
0	1	1	0
1	0	0	1
0	1	1	1

Example

Table 7.1 shows an example with four binary variables, *a*, *b*, *c*, and *d*, where each row represents a test. Of the six two-way combinations, *ab*, *ac*, *ad*, *bc*, *bd*, and *cd*, only *bd* and *cd* have all four binary values covered, so simple two-way coverage for the four tests in Table 7.1 is 2/6 = 1/3 = 33.3%. (Reading down the columns headed *c* and *d*, we see all four possible binary pairs 00, 10, 01, and 11, as is also true for *b* and *d*, but no other pair of columns contains all four.) There are four three-way combinations, *abc*, *abd*, *acd*, *bcd*, each with eight possible configurations: 000, 001, 010, 011, 100, 101, 110, 111. Of the four combinations, none has all eight configurations covered, so simple three-way coverage for this test set is 0%.

7.2.2 Simple ($t + k$)-Way

A test set for t-way interaction strength will also cover some higher-strength interactions of $t + 1$ and $t + 2$ variables.

A test set that provides full combinatorial coverage for t-way combinations will also provide some degree of coverage for ($t + 1$)-way combinations, ($t + 2$)-way combinations, and so on. This statistic may be useful for comparing two combinatorial test sets. For example, different algorithms may be used to generate three-way covering arrays. They both achieve 100% three-way coverage, but if one provides better four-way and five-way coverage, then it can be considered to provide more software testing assurance. Somewhat paradoxically, a t-way covering array that is considered "better" in the sense of containing fewer tests may end up detecting fewer faults because it covers a smaller proportion of ($t + 1$)-way combinations. More discussion of this consideration is provided in Section 9.3.

TABLE 7.2 Eight Tests for Four Binary Variables

a	*b*	*c*	*d*
0	0	0	0
1	1	1	0
1	0	0	1
0	1	1	1
0	1	0	1
1	0	1	1
1	0	1	0
0	1	0	0

Definition

For a given test set for n variables, $(t + k)$-way combination coverage is the proportion of $(t + k)$-way combinations of n variables for which all variable-value configurations are fully covered. (Note that this measure would normally be applied only to a t-way covering array, as a measure of coverage beyond t.)

Example

If the test set in Table 7.1 is extended as shown in Table 7.2, we can extend three-way coverage. For this test set, *bcd* is covered, out of the four three-way combinations, so two-way coverage is 100%, and simple $(2 + 1)$-way = three-way coverage is 25%.

7.2.3 Tuple Density

For $(t + k)$-way coverage, where $k = 1$, Chen and Zhang [41] have proposed the *tuple density* metric. A special metric for $(t + 1)$-way coverage is useful because (1) the coverage of higher-strength tuples for $t' > t + 1$ is much lower (because the number of t-way combinations to be covered grows exponentially with t), (2) the coverage at $t + 1$ provides some information for coverage at $t' > t + 1$ because $(t + 1)$-way tuples are subsumed by higher-strength tuples, and (3) the number of additional faults triggered by t-way combinations drops rapidly with $t > 2$. (This is the interaction rule.)

Definition

Tuple density is the sum of t and the percentage of the covered $(t + 1)$-tuples out of all possible $(t + 1)$-tuples [41].

Example

The test set in Table 7.2 provides 100% coverage of two-way combinations and 81% coverage of three-way combinations, so the tuple density of this test set is 2.81.

7.2.4 Variable-Value Configuration Coverage

So far, we have only considered measures of the proportion of combinations for which *all configurations* of t variables are fully covered. But when t variables with v values each are considered, each t-tuple has v^t configurations. For example, in pairwise (two-way) coverage of binary variables, every two-way combination has four configurations: 00, 01, 10, 11. We can define two measures with respect to configurations:

Definition: Variable-Value Configuration Coverage

For a given combination of t variables, variable-value configuration coverage is the proportion of variable-value configurations that are covered.

Definition: (p, t) Completeness Coverage

For a given set of n variables, (p, t)-completeness coverage is the proportion of the C(n, t) combinations that have configuration coverage of at least p [130].

Example

For Table 7.1, there are C(4, 2) = 6 possible variable combinations and $C(4, 2)2^2 = 24$ possible variable-value configurations. Of these, 19 variable-value configurations are covered and the only ones missing are $ab = 11$, $ac = 11$, $ad = 10$, $bc = 01$, $bc = 10$. But only two, *bd* and *cd*, are covered with all four value pairs (Figure 7.1). So, for the basic definition of simple t-way coverage, we have only 33% (2/6) coverage, but 79% (19/24) for the configuration coverage metric. For a better understanding of this test set, we can compute the configuration coverage for each of the six variable combinations, as shown in Figure 7.1. So, for this test set, one of the combinations (*bc*) is covered at the 50% level, three (*ab*, *ac*, *ad*) are covered at the 75% level, and two (*bd*, *cd*) are covered at the 100% level. And, as noted above, for the whole set of tests, 79% of variable-value configurations are covered. All two-way combinations have at least 50% configuration coverage, so (0.50, 2)-completeness for this set of tests is 100%.

Although the example in Table 7.1 uses variables with the same number of values, this is not a requirement. Coverage measurement tools that we have developed compute coverage for test sets in which parameters may have differing numbers of values, as shown in Figure 7.7.

Figure 7.2 shows a graphical display of the coverage data for the tests in Table 7.1. Coverage is given as the Y axis, with the percentage

Vars	Configurations	Config
a b	00, 01, 10	0.75
a c	00, 01, 10	0.75
a d	00, 01, 11	0.75
b c	00, 11	0.50
b d	00, 01, 10, 11	1.0
c d	00, 01, 10, 11	1.0

- Total 2-way coverage = 19/24 = 0.79167
- (0.50, 2)-completeness = 6/6 = 1.0
- (0.75, 2)-completeness = 5/6 = 0.83333
- (1.0, 2)-completeness = 2/6 = 0.33333

FIGURE 7.1 The test array covers all possible two-way combinations of *a*, *b*, *c*, and *d* to different levels.

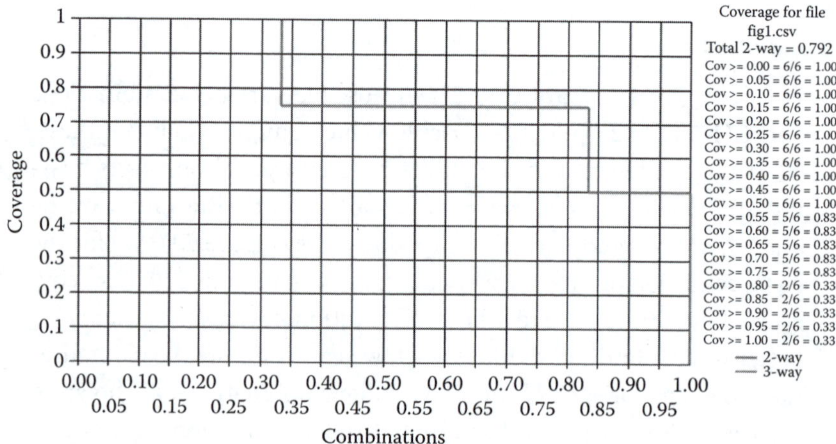

FIGURE 7.2 Graph of coverage from example test data above.

of combinations reaching a particular coverage level as the X axis. Note from Figure 7.1 that 100% of the combinations are covered to at least the 0.50 level, 83% are covered to the 0.75 level or higher, and a third covered 100%. Thus, the rightmost horizontal line on the graph corresponds to the smallest coverage value from the test set, in this case 50%. Thus, (.50, 2)-completeness = 100%.

Note that the total two-way coverage is shown as 0.792. This figure corresponds approximately to the area under the curve in Figure 7.2. (The area is not exactly equal to the true figure of 0.792 because the curve is plotted in increments of 0.05.) Counting the squares and partial squares, we see that there are approximately 40 squares in the upper right corner of the graph, so the area below the curve is roughly 160/200. Additional tests can be added to provide greater coverage. Suppose an additional test [1,1,0,1] is added. Coverage then increases as shown in Figure 7.3. Now we can see immediately that all combinations are covered to at least the 75% level, as shown by the rightmost horizontal line. (In this case, there is only one horizontal line.) The leftmost vertical line reaches the 1.0 level of coverage, showing that 50% of combinations are covered to the 100% level.

Note that the upper right corner includes roughly 25 squares, so the area under the curve is 175/200, or 87.5%, matching the total two-way coverage figure. Adding yet another test, [1,0,1,1] results in the coverage shown in Figure 7.4.

Finally, if a test [1,0,1,0] is then added, 100% coverage for all combinations is reached, as shown in Figure 7.5. The graph can thus be thought of

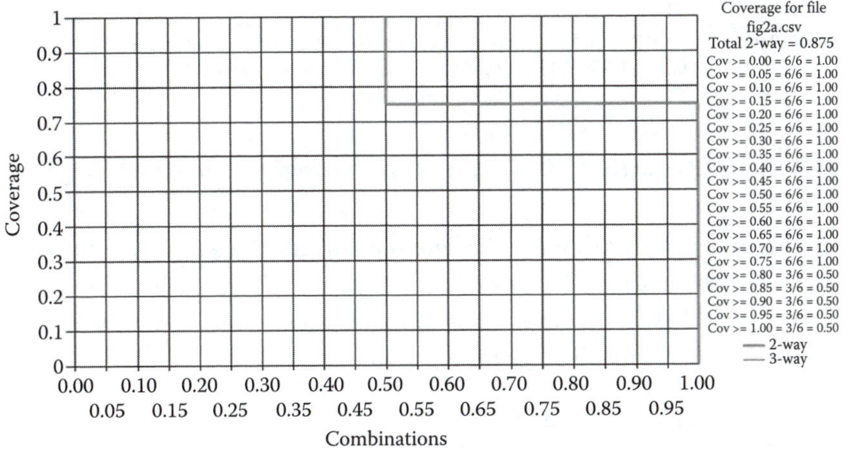

FIGURE 7.3 Coverage after test [1,1,0,1] is added.

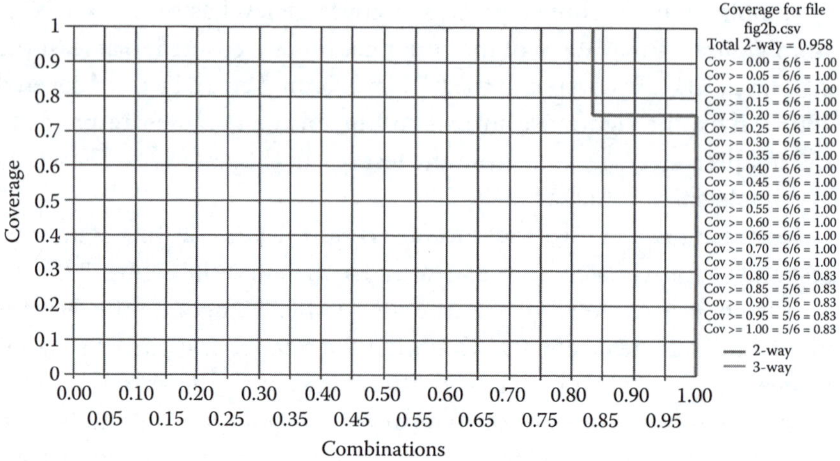

FIGURE 7.4 Coverage after test [1,0,1,1] is added.

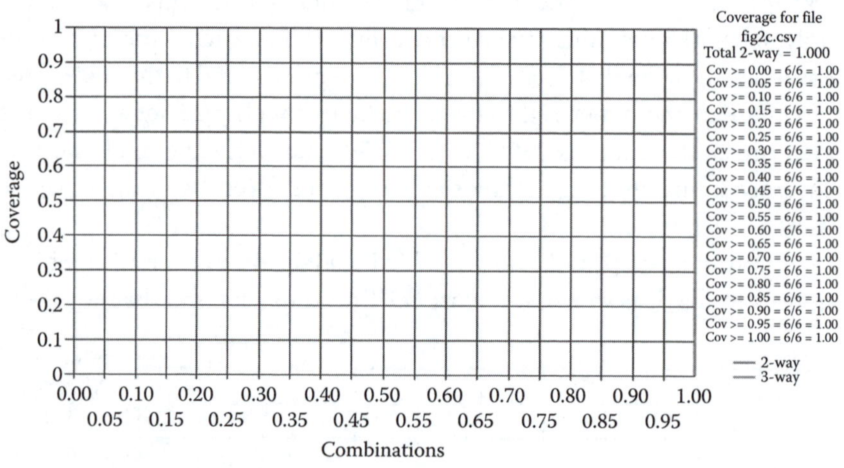

FIGURE 7.5 Adding test [1,0,1,0] provides 100% coverage, that is, a covering array.

as a "coverage strength meter," where the indicator line moves to the right as higher coverage levels are reached. A covering array, which by definition covers 100% of variable-value configurations, will always appear as in Figure 7.5, with full coverage for all combinations.

7.3 USING COMBINATORIAL COVERAGE

What is the point of the measures introduced in the previous section? How can they help with testing? There are two important applications of coverage measures. One is to get a better understanding of how effective a test suite may be if it was not designed as a covering array. If we have a test suite developed by analyzing requirements and constructing tests *ad hoc*, then coverage measurement may suggest interactions that are not being tested thoroughly. Another use is to measure important characteristics of $(t + 1)$-way and higher-strength coverage of a t-way test suite.

We know that any t-way covering array inevitably includes higher-strength combinations, and we need ways to describe this additional test strength. One is the tuple density metric, introduced previously. For example, the example in Table 7.2 was shown to have a tuple density of 2.81, because it covers all two-way combinations and 81% of three-way combinations. But consider what happens in applying the array in Table 7.2 to a testing problem. We know it will detect two-way faults but how likely is it that the test suite will also detect a particular three-way fault? Even though it fully covers 25% of three-way combinations, the likelihood of detecting a particular three-way fault is in fact much higher than 25%. It may seem surprising at first, but in fact this array covers more than 80% of the three-way variable-value configurations, as shown in Figure 7.6. Note that for four binary variables, there are $2^3C(4, 3) = 32$ three-way variable-value configurations. Analyzing the array, we find that it covers 26 of the 32, or 81.25%. Even though only one of the four three-way combinations is completely covered with all $2^3 = 8$ possible settings, the other three-way combinations have most of the eight settings covered, so that

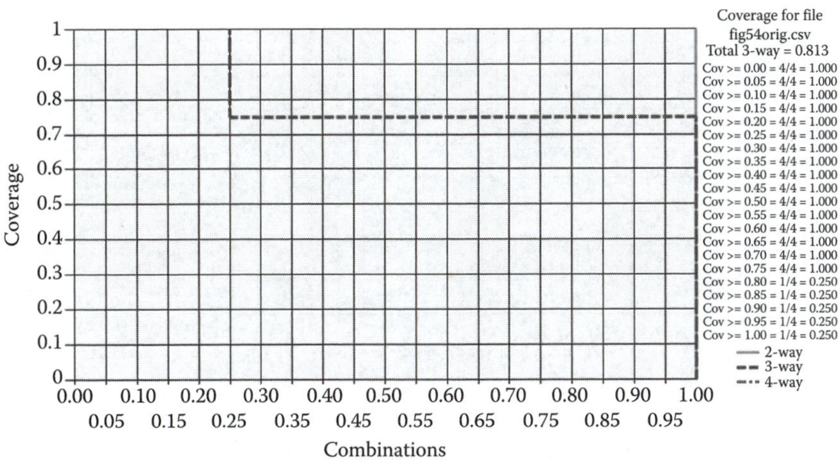

FIGURE 7.6 Coverage of tests in Table 7.2.

the total number of variable-value configuration settings covered is 26. Since it covers 81% of the three-way combinations, this particular two-way covering array may detect three-way faults as well.

The methods described in this chapter were originally developed to analyze the state space coverage of spacecraft software [130]. As with many assurance efforts, a very thorough set of over 7000 tests had been developed for each of the three systems, but combinatorial coverage was not the goal. With such a large test suite, it seemed likely that a huge number of combinations had been covered, but how many? Did these tests provide two-way, three-way, or even higher-degree coverage? If an existing test suite is relatively thorough, it may be practical to supplement it with a few additional tests to bring coverage up to the desired level.

The original test suites had been developed to verify correct system behavior in normal operation as well as a variety of fault scenarios, and performance tests were also included. Careful analysis and engineering judgment were used to prepare the original tests, but the test suite was not designed according to criteria such as statement or branch coverage. The system was relatively large, with 82 variables in a $1^3 2^{75} 4^2 6^2$ configuration (three 1-value, 72 binary, two 4-value, and two 6-value). Figure 7.7 shows combinatorial coverage for this system. This particular test set was not a covering array, but pairwise coverage is still relatively good, with about

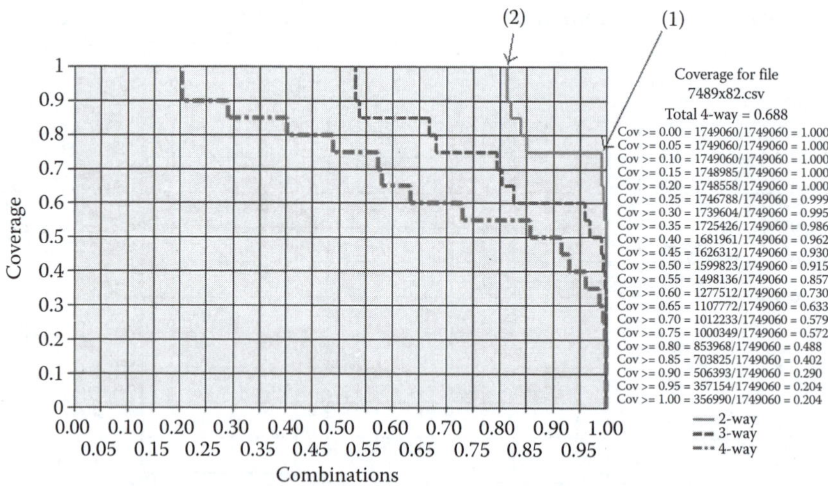

FIGURE 7.7 **(See color insert.)** Configuration coverage for $1^3 2^{75} 4^2 6^2$ inputs. (1) Configurations covered to 75% level and (2) configurations covered to 100% level.

99% of the two-way combinations having at least 75% of possible variable-value configurations covered (1), and 82% have 100% of possible variable-value configurations covered (2).

Consider the graph in Figure 7.8. The label Φ indicates the percentage of combinations with 100% variable-value coverage, and *M* indicates the percentage of coverage for 100% of the *t*-way combinations. In this case, 33% (Φ) of the combinations have full variable-value coverage, and all combinations are covered to at least the 50% level (*M*). So, (1.0, 2)-completeness = Φ and (*M*, 2)-completeness = 100%, but it is helpful to have more intuitive terms for the points Φ and *M*. Note that Φ is the level of simple *t*-way coverage. Since all combinations are covered to at least the level of *M*, we will refer to *M* as the *t-way minimum coverage*, keeping in mind that "coverage" refers to a proportion of variable-value configuration values. Where the value of *t* is not clear from the context, these measures are designated Φ_t and M_t. Using these terms, we can analyze the relationship between total variable-value configuration coverage, *t*-way minimum coverage, and simple *t*-way coverage.

Let S_t be total variable-value coverage, the proportion of variable-value configurations that are covered by at least one test. If the area of the entire graph is 1 (i.e., 100% of combinations), then

$$S_t \geq 1 - (1-\Phi_t)(1-M_t)$$
$$S_t \geq \Phi_t + M_t - \Phi_t M_t \qquad (7.1)$$

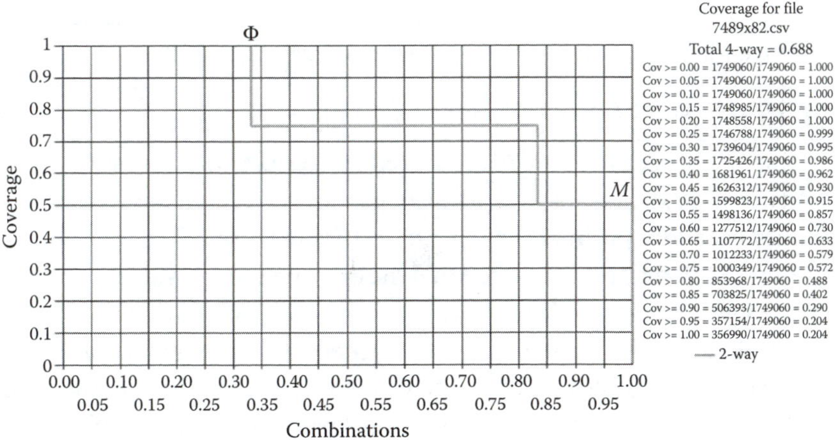

FIGURE 7.8 Example coverage graph for $t = 2$.

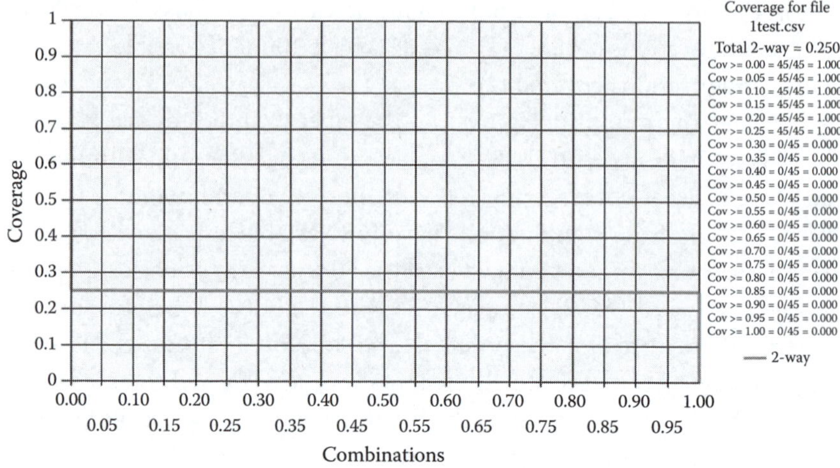

FIGURE 7.9 Coverage for one test, 10 binary variables.

If a test suite has only one test, then it covers C(n, t) combinations. The total number of combinations that must be covered is C(n, t) × v^t, so the coverage of one test is $1/v^t$. Thus

$$t\text{-way minimum coverage}: M_t \geq \frac{1}{v^t} > 0 \quad (7.2)$$

Thus, for any nonempty test suite, t-way minimum coverage ≥ $1/v^t$. This can be seen in Figure 7.9, which shows coverage for a single test with binary variables, where two-way minimum coverage is 25%. With only one test, two-way full coverage is of course 0, and S = total variable-value coverage (denoted "total t-way" on the graph legend) is 25%.

7.4 COST AND PRACTICAL CONSIDERATIONS

An important cost advantage introduced by coverage measurement is the ability to use existing test sets, identify particular combinations that may be missing, and supplement existing tests. In some cases, as in the example of Figure 7.7, it may be discovered that the existing test set is already strong with respect to a particular strength t (in this case two-way), and tests for $t + 1$ generated. The trade-off in cost of applying coverage measurement is the need to map existing tests into discrete values that can be analyzed by the coverage measurement tools (see Appendix C). For example, if the test set includes a field with various dollar amounts, it may be necessary to establish equivalence classes such as 0, 1,. . .,1000, >1000, and so on.

These concepts can be used to analyze various testing strategies by measuring the combinatorial coverage they provide.

All-values: Consider the t-way coverage from one of the most basic test criteria, all-values, also called "each-choice" [2]. This strategy requires that every parameter value be covered at least once. If all n parameters have the same number of values, v, then only v tests are needed to cover all. Test 1 has all parameters set to v_1, Test 2 to v_2, and so on. If parameters have different numbers of values, where parameters $p_1, \ldots, p_n$ have v_i values each, the number of tests required is at least $Max_{i=1,n}\ v_i$.

Example: If there are three values, 0, 1, and 2, for five parameters, then 0,0,0,0,0; 1,1,1,1,1; and 2,2,2,2,2 will test all values once. Since each test covers $1/v^t$ variable-value configurations, and no combination appears in more than one test, with v values per parameter, we have

$$M_t(all\text{-}values) \geq v\frac{1}{v^t} = \frac{1}{v^{t-1}}$$

So, if all values are covered at least once, for all-values, minimum coverage $M \geq 1/v^{t-1}$. This relationship can be seen in Figure 7.10, which shows coverage for two tests with binary variables; two-way minimum coverage = 0.5, and three-way and four-way coverage are 0.25 and 0.125, respectively.

Base choice: Base-choice testing [3] requires that every parameter value be covered at least once and in a test in which all the other values are

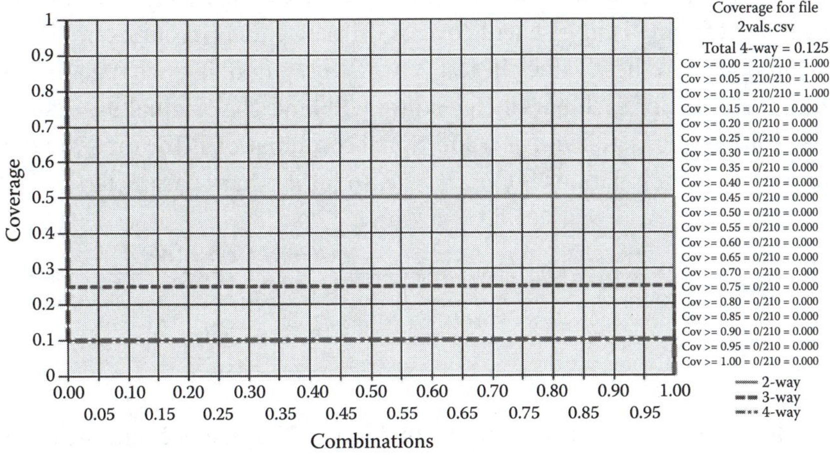

FIGURE 7.10 **(See color insert.)** t-Way coverage for two tests with binary values.

	a	b	c	d
test 1	0	0	0	0
test 2	1	0	0	0
test 3	0	1	0	0
test 4	0	0	1	0
test 5	0	0	0	1

FIGURE 7.11 Base choice test suite for a 2^4 configuration, where test 1 is base choice values.

held constant. Each parameter has one or more values designated as base choices. The base choices can be arbitrary, but can also be selected as "more important" values, for example, default values, or values that are used most often in operation. If parameters have different numbers of values, where parameters $p_1, \ldots, p_n$ have v_i values each, the number of tests required is at least $1 + \Sigma_{i=1,n} (v_i - 1)$, or where all n parameters have the same number of values v, the number of tests is $1 + n(v - 1)$. An example is shown in Figure 7.11, with four binary parameters.

The base choice strategy can be highly effective, despite its simplicity. In one study of five programs seeded with 128 faults [91,92], it was found that although the Base Choice strategy requires fewer test cases than some other approaches, it found as many faults [76]. In that study, AETG [61] was used to generate two-way (pairwise) test arrays. We can use combinatorial coverage measurement to help understand this finding. For this example of analyzing base choice, we will consider n parameters with two values each. First, note that the base choice test covers C(n, t) combinations, so for pairwise testing, this is C(n, 2) = $n(n - 1)/2$. Changing a single value of the base test to something else will cover $n - 1$ new pairs. (In our example, *ab*, *ac*, and *ad* have new values in test 2, while *bc*, *cd* are is unchanged.) This must be done for each parameter, so we will have the original base choice test combinations plus $n(n - 1)$ additional combinations. The total number of two-way combinations is C(n, 2) × 2^2, so for n binary parameters:

$$M_t(\textit{two-way binary base choice}) = \frac{n(n-1)/2 + n(n-1)}{C(n,2)2^2}$$

$$= \frac{C(n,2) + 2C(n,2)}{C(n,2)2^2} = 3/4$$

This can be seen in the graph in Figure 7.12 of coverage measurement for the test suite in Figure 7.11. Note that the 75% coverage level is independent of n. For $v > 2$, the analysis is a little more complicated. The

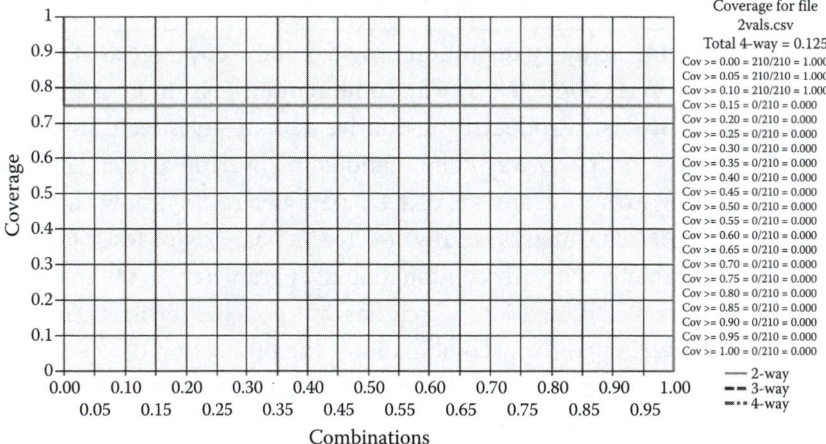

FIGURE 7.12 Graph of two-way coverage for test set in Figure 7.11.

base choice test covers C(n, 2), and each new test covers $n - 1$ new combinations, but we need $v - 1$ additional tests beyond the base choice for v values. So, the proportion of two-way combinations covered by the base choice strategy in general is

$$M_t(base\ choice) = \frac{C(n,2) + 2(v-1)C(n,2)}{C(n,2)v^2} = \frac{1+2(v-1)}{v^2}$$

For example, with $v = 3$ values per parameter, we still cover 55.6% of combinations, and with $v = 4$, coverage is 43.75%. This analysis helps explain why the base choice strategy compares favorably with pairwise testing.

This equation can be generalized to higher interaction strengths. The base choice test covers C(n, t) combinations, and each new test covers C($n - 1$, $t - 1$) new combinations, since the parameter value being varied can be combined with $t - 1$ other parameters for each t-way combination. Base choice requires $n(v - 1)$ additional tests beyond the initial one, so for any n, t with $n \geq t$

$$M_t\ (base\ choice) = \frac{C(n,t) + n(v-1)C(n-1,t-1)}{C(n,t)v^t}$$

$$= \frac{C(n,t) + t(v-1)C(n,t)}{C(n,t)v^t} = \frac{1+t(v-1)}{v^t}$$

7.5 ANALYSIS OF $(t+1)$-WAY COVERAGE

A t-way covering array by definition provides 100% coverage at strength t, but it also covers some $(t+1)$-way combinations (assuming $n \geq t+1$). Using the concepts introduced previously, we can investigate minimal coverage levels for $(t+1)$-way combinations in covering arrays. Given a t-way covering array, for any set of $t+1$ parameters, we know that any combination of t parameters is fully covered in some set of tests. Joining any other parameter with any combination of t parameters in the tests will give a $(t+1)$-way combination, which has v^{t+1} possible settings. For any set of tests covering all t-way combinations, the proportion of $(t+1)$-way combinations covered is thus v^t/v^{t+1}, so if we designate $(t+1)$-way variable-value configuration coverage as S_{t+1}, then

$$S_{t+1} \geq \frac{1}{v} \quad \text{for any } t\text{-way covering array with } n \geq t+1 \qquad (7.3)$$

Note that variable-value configuration coverage is not the same as simple t-way coverage, which gives the proportion of t-way configurations that are 100% covered. Clearly, no set of $(t+1)$-way combinations will be fully covered with a t-way covering array, if $N < v^{t+1}$, where $N =$ number of tests. For most levels of t and v encountered in practical testing, this condition will hold. For example, if $v = 3$, then a two-way covering array with $<3^3 = 27$ tests can be computed [71] for any test problem with <60 parameters. So, the proportion of combinations with full variable-value coverage, designated Φ, will be zero for three-way coverage for this example. And, in general, designating $(t+1)$-way full variable-value configuration coverage as Φ_{t+1}, if $N < v^{t+1}$, then $\Phi_{t+1} = 0$, for any t-way covering array with $n \geq t+1$.

As an additional illustration, we show that expression (7.3) can also be reached by noting that with a t-way covering array, unique $(t+1)$-way combinations can be identified as follows: traversing the covering array, for the first appearance of each t-way combination, appends each of the $n-t$ parameters not contained in the t-way combination to create a $(t+1)$-way combination. This procedure counts each $(t+1)$-way combination $C(t+1, t) = t+1$ times. The covering array contains $C(n, t)v^t$ t-way combinations, so the proportion of $(t+1)$-way combinations, S_{t+1}, in the t-way array is at least

$$\frac{C(n,t)v^t(n-t)/(t+1)}{C(n,t+1)v^{t+1}} = \frac{1}{v}$$

In most practical cases, the $(t + 1)$-way coverage of a t-way array will be higher than $1/v$.

7.6 CHAPTER SUMMARY

1. Many coverage measures have been devised for code coverage, including statement, branch or decision, condition, and modified condition decision coverage. These measures are based on aspects of source code and are not suitable for combinatorial coverage measurement.

2. Measuring configuration-spanning coverage can be helpful in understanding state space coverage. If we do use combinatorial testing, then configuration-spanning coverage will be 100% for the level of t that was selected, but we may still want to investigate the coverage our test set provides for $t + 1$ or $t + 2$. Calculating this statistic can help in choosing between t-way covering arrays generated by different algorithms. As seen in the examples above, it may be relatively easy to produce tests that provide a high degree of spanning coverage, even if not 100%. In many cases, it may be possible to generate additional tests to boost the coverage of a test set.

3. The following properties hold:

$$M_t \geq \frac{1}{v^t} > 0$$

$$S_t \geq \Phi_t + M_t - \Phi_t M_t$$

$$S_{t+1} \geq \frac{1}{v} \quad \text{for any } t\text{-way covering array with } n \geq t + 1$$

If $N < v^{t+1}$, then $\Phi_{t+1} = 0$, for any t-way covering array with $n \geq t + 1$.

REVIEW

Q: What is the key difference between combinatorial coverage and traditional measures such as statement, decision, or condition coverage?

A: Combinatorial coverage includes measures of the input state space, while the other coverage measures listed deal with paths through a program.

Q: What is simple t-way combination coverage?

A: It is the proportion of t-way combinations of n variables for which all variable-value configurations are fully covered.

Q: What is variable-value configuration coverage?

A: Variable-value configuration coverage is the proportion of variable-value configurations that are covered.

Q: What is (p, t)-completeness?

A: (p, t)-completeness is the proportion of the $C(n, t)$ combinations that have variable-value configuration coverage of at least p.

Q: Consider the square in the upper right corner of Figure 7.4, which is approximately $3.3 \times 2.5 = 8.25$ blocks in size. The complete graph contains 200 blocks, so the upper right corner is approximately 4% of the total graph. Why?

A: Total coverage is shown in the graph to be 95.8%, so the part in the upper right corner corresponds to the proportion of combinations not covered. Because the graph is constructed with variable-value configuration coverage statistics grouped into 20 bins of 5% each, the proportions of covered/uncovered values will not match the graph precisely, but will be close.

Q: Using Colbourn's covering array size tables, determine the number of tests for a two-way covering array for a test problem with at least 10 parameters, all of which have three or more values. How does this size compare with the number of tests required for the base choice strategy? Colbourn's tables can be found at: http://www.public.asu.edu/~ccolbou/src/tabby/catable.html.

A: For example, 20 parameters with three values each will require a two-way covering array of 15 tests. For base choice, we would need $1 + 20(3 - 1) = 41$ tests.

Q: Why is the covering array so much smaller in the previous example?

A: For a given value of t, the size of the covering array grows proportional to $\log n$, but for base choice, the number of tests increases proportional to n.

Q: How does combinatorial testing compare with base choice for strengths higher than pairwise?

A: The number of tests in a covering array grows proportional to v^t, so for $t > 2$, the covering array may be much larger than the base choice test set size.

CHAPTER 8

Test Suite Prioritization by Combinatorial Coverage

Renée Bryce and Sreedevi Sampath

The benefits of combinatorial testing extend beyond that of test suite generation. Several research studies have shown the benefits of using combinatorial-based coverage for test suite prioritization—that is, how can we arrange tests in the most effective order to reduce cost and speed up fault detection? This chapter reviews combinatorial prioritization criteria and algorithms and presents a freely available tool for researchers and practitioners to use in their own work.

8.1 COMBINATORIAL COVERAGE FOR TEST SUITE PRIORITIZATION

Previous chapters introduced combinatorial testing and many studies that show the benefits of this technique in different domains. This chapter presents a different view of combinatorial testing for the purpose of test suite prioritization that has been applied in the GUI and web testing domains. In the remainder of this section, we will review the test suite prioritization problem, inter-window combinatorial coverage for event-driven software, and an example.

Test efficiency can be increased by ordering tests according to combinatorial coverage.

Regression testing occurs during the maintenance phase of the software development lifecycle. Typically, a large number of test cases accumulate over several lifecycles of the software system. It is often not practical to use all test cases when testing a new version of a system, because test cases can be too numerous, redundant, or obsolete. In test suite prioritization, test cases in a test suite are ordered based on their importance according to some criteria to meet a performance goal. An example performance goal is to find faults as quickly as possible in the test execution cycle. Formally, the *test suite prioritization problem* as defined by Rothermel et al. [167] is stated as follows:

> Given (T, Π, *f*), where T is a test suite, Π is the set of all test suites that are prioritized orderings of T obtained by permuting the tests of T, and *f* is a function to evaluate the orderings from Π to the real numbers, the problem is to find a permutation, $\pi \in \Pi$ such that $\forall \pi' \in \Pi, f(\pi) \geq f(\pi')$.

For event-driven software, such as GUIs and web applications [28–30,173], existing work has used *inter-window combinatorial coverage* to prioritize the test suites. The measure inter-window combinatorial coverage is defined as the number of combinations of parameter values on two different windows.

Figures 8.1 and 8.2 show two pages from a web-based application, SchoolMate, which is a web software system designed to manage a school's classes, students, and teachers. Tables 8.1 and 8.2 list the parameters and possible values for each of these pages, and Table 8.3 gives the parameters on different windows. Imagine a logic fault where the developer mistakenly makes it impossible to generate a grade report if a student has an absence. This is triggered by the combination of a user *clicking "Add Attendance"* with the default value of *"Tardy"* in the *"Type"* dropdown box on the first page shown in Figure 8.1 in combination with choosing *"Grade Report"* for the *"Choose a Report"* parameter shown in Figure 8.2. Specifically, the 2-way combination that triggers the fault is Page 1: (Parameter = Add Attendance, Value = left click) and Page 2: (Parameter = Choose a Report, Value = Grade Report). Faults of this nature that are detected by

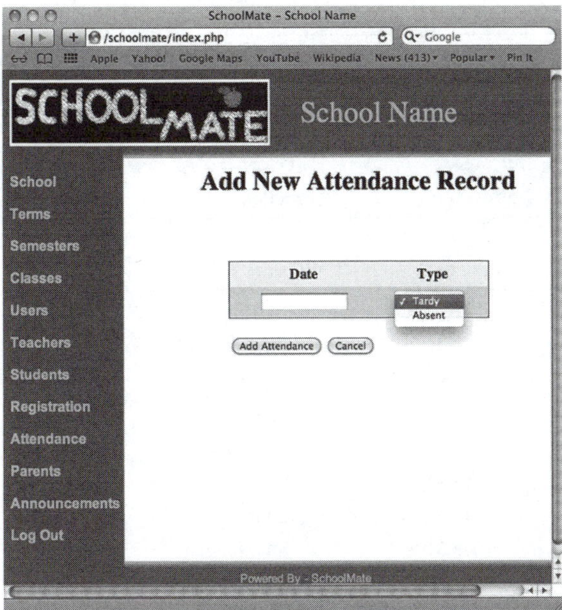

FIGURE 8.1 **(See color insert.)** Screenshot of the SchoolMate page for "Add New Attendance Record."

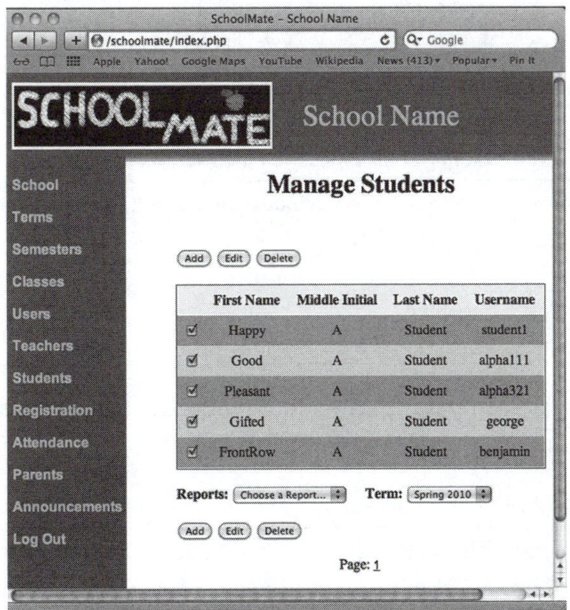

FIGURE 8.2 **(See color insert.)** Screenshot of the SchoolMate page for "Manage Students."

TABLE 8.1 Parameters and Values from Figure 8.1

Parameter	Type	Possible Valid Values	Number of Possible Values
Date	Text box	Any text	1..*
Type	Dropdown box	Tardy, Absent	2
Add attendance	Button	Left click	1
Cancel	Button	Left click	1

TABLE 8.2 Parameters and Values from Figure 8.2

Parameter	Type	Possible Valid Values	Number of Possible Values
Add	Button	Left click	1
Edit	Button	Left click	1
Delete	Button	Left click	1
Checkbox1	Checkbox	Enable, disable	2
Checkbox2	Checkbox	Enable, disable	2
Checkbox3	Checkbox	Enable, disable	2
Checkbox4	Checkbox	Enable, disable	2
Checkbox5	Checkbox	Enable, disable	2
Choose a Report	Dropdown box	Deficiency Report, Grade Report, Points Report, Report Cards	4
Term	Dropdown box	Spring 2010	1
Add	Button	Left click	1
Cancel	Button	Left click	1

TABLE 8.3 Example Web Application with Windows and Parameters

Window	Parameters on Window That Can Be Set to Values
W1	p1, p2, p3
W2	p4, p5
W3	p6, p7, p8
W4	p9

interactions between parameters will be detected by test cases that have such inter-window combinatorial coverage.

8.2 ALGORITHM

The algorithm to prioritize a test suite, T, uses a simple greedy process [25,28,172]. Figure 8.3 shows the pseudocode for this process. Begin with an unordered test suite, T, and then iteratively select the test case that covers the most previously uncovered *t*-tuples to add one test case at a time to the prioritized test suite, PT. The variable, "bestTest" denotes the test case

```
Prioritize(TestSuite T)
        TestSuite PT
        maxTuplesCovered=-1
        bestTest = 0
        for (i=0; i< sizeOf(T); i++) {
          for (j=0; j<sizeOf(T); j++){
            if(testUsed(j) == false){
              tuplesCovered = computeTuplesCovered(t_j)
              if(maxTuplesCovered < tuplesCovered){
                maxTuplesCovered = tuplesCovered
                bestTest = i
              }
              else if(maxTuplesCovered == tuplesCovered){
                  breakTieAtRandom(i);
              }
            }
          }
          add bestTest to PT
          markTuplesCovered(bestTest)
          markBestTestAsUsed(bestTest)
        }
        return(PT)
}
```

FIGURE 8.3 Pseudocode for the prioritization algorithm.

that covers the most previously uncovered *t*-tuples. When this test case is added to the prioritized test suite, PT, the algorithm marks this test case as used and marks the *t*-tuples from this test case as covered since future tests do not receive preference for covering these *t*-tuples again. This process continues until all of the tests from the original test suite are added to the prioritized test suite. Ties are broken at random.

8.3 PRIORITIZATION CRITERIA

Two combinatorial prioritization criteria, 1-way and 2-way, are defined and experimentally evaluated in the literature [28,180]. The 1-way criterion selects test cases in nonascending order of the number of parameter values in each test case.

Consider the example test suite shown in Table 8.4 for the example web application whose windows and parameters are shown in Table 8.3.

The 1-way score for T1 is 4, since it covers four parameters, T2 is 5, T3 is 3, and T4 is 4. The test cases will be ordered based on the number of parameter values in them. First, T2 is selected, and all the parameters covered by T2, that is, parameters p4, p5, p2, p3, p9, are marked as selected. Thus, the 1-way score of the remaining test cases is as follows: T1 is 2, T3 is 1, and T4 is 1. From the remaining test cases, T1 is selected next since it covers the most parameters. Marking off these parameters as covered in the remaining test results in the update of the 1-way score as follows:

TABLE 8.4 Example Test Suite for Web Application

Test Case	Windows Visited	Parameter Set to Values
T1	W1 -> W2 -> W3	p1, p2, p4, p6
T2	W2 -> W1 -> W4	p4, p5, p2, p3, p9
T3	W3 -> W2 -> W4	p7, p5, p9
T4	W4 -> W1	p9, p1, p2, p3

T3 is 1, T4 is 0. The next test to be selected is thus T3 followed by T4. The prioritized test suite by the 1-way criterion contains the test cases in the following order: {T2, T1, T3, T4}.

The two-way pairs covered by the test cases from Table 8.4 are shown in Table 8.5. As we can see from Table 8.5, only pairs of parameters *between windows* are considered when computing the pairs for the 2-way criterion. For example, for test case T1, we consider pairs of parameters between p1 and p4, p1 and p6, but not between p1 and p2, because both p1 and p2 are parameters on the same window, W1 (see Table 8.3). The intuition behind computing pairs in this manner is that parameters set to values between windows are likely to expose more faults in event-driven software. Future work may examine other combinations.

Using Table 8.5, we can create the prioritized test suite by the 2-way criterion. First, we select the test case that covers the most 2-way pairs, which in this example is T2. Then, we mark all the pairs covered by T2 as covered in the other test cases. This updates the 2-way score of each test case as follows: T1's score is 4 (because (p4, p2) is covered by T2), T4's score is 1 (since (p9, p2) is covered by T2), and T3's score is 2 (since the pair (p5, p9) is covered by T2). Note that order between pairs is not considered in this work, that is, the pair (p4, p2) is considered to be the same as the pair (p2, p4). In the next iteration of the prioritization algorithm, we select T1 since it covers the most uncovered pairs and mark off all the pairs covered by T1 as covered in the other tests. The 2-way scores are updated as follows: T3's and T4's scores are unchanged because the pairs covered by T1 do not appear in T3 or T4. Now, since both T3 and T4 cover the same number of

TABLE 8.5 2-Way Pairs Covered by the Test Cases

Test Case	2-Way Pairs of Parameters Covered
T1	(p1, p4), (p1, p6), (p2, p4), (p2, p6), (p4, p6)
T2	(p4, p2), (p4, p3), (p4, p9), (p5, p2), (p5, p3), (p5, p9), (p2, p9), (p3, p9)
T3	(p7, p5), (p7, p9), (p5, p9)
T4	(p9, p1), (p9, p2), (p9, p3)

pairs, the selection between them is done at random. The final prioritized test suite by 2-way criterion is {T2, T1, T4, T3}.

8.4 REVIEW OF EMPIRICAL STUDIES

The following results are from a subset of the empirical studies presented in Bryce et al. [28] and Sherwood et al. [180], which show the effectiveness of prioritization. The empirical studies were conducted on event-driven software, such as GUIs and web applications. Here, we present the results of the studies conducted on web applications.

8.4.1 Subject Applications

Three web applications and user-session-based test suites were used in the experiments. The applications are nontrivial as most contain several thousand lines of code, from 9 to 219 classes, 22 to 644 methods, and from 108 to 1720 branches. The test cases exercise from 85 to 4166 inter-window parameter values on pages of these applications. The first application, Book, allows users to register, login, browse for books, search for books by keyword, rate books, add books to a shopping cart, modify personal information, and logout. Only functionality accessible to the end-user/consumer was included in the study. Book uses JSP for its front-end and a MySQL database back-end. The second application, Course Project Manager (CPM), allows course instructors to login and create grader accounts for teaching assistants. Instructors and teaching assistants create group accounts for students, assign grades, and create schedules for demonstration time slots. Users interact with an HTML application interface generated by Java servlets and JSPs. The third application, the conference management system for the Mid-Atlantic Symposium on Programming Languages And Systems (MASPLAS) enables users to register for a workshop, upload abstracts and papers, and view the schedule, proceedings, and other related information. MASPLAS is written using Java, JSP, and MySQL.

8.4.2 Prioritization Criteria

We present five of the prioritization criteria that were evaluated in the study [28]:

- *1-Way:* select tests in nonascending order of the number of parameter values in each test case.
- *2-Way:* select tests in nonascending order of the number of pairwise inter-window parameter-value interactions in each test case.

- *Random:* Randomly order the test cases (using the rand() function).
- *G-Best:* Using the known fault matrix, greedily order the test cases in nonascending order of the number of faults that each test case identifies.
- *G-Worst:* Using the known fault matrix, greedily order the test cases in nondescending order of the number of faults that each test case identifies.

8.4.2.1 Test Cases

The test cases used in this study were generated from field usage data. The web applications were deployed to a limited user group. All user accesses were recorded by a logger on the web server. The user accesses were in the form of HTTP GET and POST requests with associated data. The requests were then converted into test cases based on cookies, the IP address from where the request originated, and a timeout interval of 45 min [180]. This led to the creation of 125 test cases for Book, 890 test cases for CPM, and 169 test cases for MASPLAS. Table 8.6 shows some characteristics of the test cases.

8.4.2.2 Faults

To perform the controlled experiment, faults were manually seeded into the application. The different types of faults seeded in the application are: logic faults (faults in application code logic and flow), appearance faults (faults that affect the way the page is displayed to the user), data store faults (faults that affect the code that interacts with the data store), form faults (faults in HTML form fields and actions), and link faults (faults that alter the location of hyperlinks). Table 8.6 also shows the number of faults seeded in each application and the average number of faults detected by a test case.

TABLE 8.6 Characteristics of Test Cases and Faults

	Book	CPM	MASPLAS
Total number of tests	125	890	169
Number of windows in largest test case	160	585	69
Average number of windows in a test case	29	14	7
Number of seeded faults	40	135	29
Average number of faults found by a test case	21.43	4.67	4.62

8.4.2.3 Evaluation Metric

The evaluation metric Average Percent of Faults Detected (APFD) is used to measure the effectiveness of the prioritization criteria. APFD is a measure of the rate at which faults are detected. Informally, it is the area under the curve in which the x-axis represents each test case and the y-axis represents the number of faults detected. Formally, Rothermel et al. [167] defined APFD as follows: For a test suite, T with n test cases, a set of faults, F with m faults detected by T, then let TF_i be the position of the first test case t in T′, where T′ is an ordering of T, that detects fault i. Then, the APFD metric for T′is as follows:

$$\text{APFD} = 1 - \frac{\text{TF}_1 + \text{TF}_2 + \cdots + \text{TF}_m}{mn} + \frac{1}{2n}$$

8.4.2.4 Results

Tables 8.7 through 8.9 show the APFD results for Book, CPM, and MASPLAS in 10% increments of the test suite. Criteria with higher APFD values are better than criteria with lower APFD values since they indicate that the faults are found early in the test execution cycle using the test order generated by the criterion. The bold-face values in the tables represent the highest APFD value for that portion of the test suite.

TABLE 8.7 APFD for Book

	10%	20%	30%	40%	50%	60%	70%	80%	90%	100%
1-Way	**93.44**	**93.44**	**93.44**	**93.44**	**94.96**	**96.13**	**96.13**	**96.13**	**96.13**	**96.13**
2-Way	93.22	93.22	93.22	93.22	94.69	94.69	95.62	95.62	95.62	95.62
Random	86.36	93.7	94.52	94.86	84.86	95.11	95.27	95.56	95.56	95.57
G-Best	94.02	96.38	96.38	96.38	96.38	96.38	96.38	96.38	96.38	96.89
G-Worst	53.73	71.91	84.82	84.82	82.82	84.82	85.82	85.82	85.82	85.87

TABLE 8.8 APFD for CPM

	10%	20%	30%	40%	50%	60%	70%	80%	90%	100%
1-Way	**83.79**	87.78	91.54	94.79	94.79	94.79	94.79	94.79	94.99	94.99
2-Way	83.72	**90.8**	**91.72**	**95.64**	**95.64**	**95.64**	**95.64**	**95.64**	**95.64**	**95.64**
Random	48.63	57.55	64.51	69.19	73.03	75.37	77.37	78.24	78.45	78.49
G-Best	95.27	97.37	97.37	97.37	97.37	97.92	97.92	97.92	97.92	97.92
G-Worst	9.14	11.18	12.97	29.96	27.6	30.99	30.99	30.99	33.15	34.64

TABLE 8.9 APFD for MASPLAS

	10%	**20%**	**30%**	**40%**	**50%**	**60%**	**70%**	**80%**	**90%**	**100%**
1-Way	89.6	**93.04**	93.04	95.56	95.56	95.56	95.56	95.56	95.56	95.56
2-Way	**90.98**	90.98	**94.28**	**97.06**	**97.06**	**97.06**	**97.06**	**97.06**	**97.06**	**97.06**
Random	76.33	80.51	85.57	87.59	89.91	90.69	90.69	90.91	90.91	90.91
G-Best	99.5	100	100	100	100	100	100	100	100	100
G-Worst	87.15	87.15	87.15	89.22	89.81	90.85	92.06	92.18	92.18	92.28

We see that the criteria of 1-way and 2-way perform better than Random. In Book [171], Random performs comparable to 2-way but that is an artifact of the test suite and faults seeded in Book. From Table 8.6, we see that the average faults detected by a test case are 21 for Book, while around 4 for CPM and MASPLAS. This indicates that the test cases in Book, on average, detect a large number of the seeded faults. Therefore, Random selection of test cases is likely to perform well because any test case that is selected at random may be detecting a large number of faults.

The 2-way test criterion performed close to the theoretical best case scenario.

The G-Best is a theoretical best-case scenario and thus will always have the best APFD value. However, it is important to note that the 1- and 2-way APFD values are close to the theoretical best-case scenario. For Book, we see that 1-way criterion is better than the 2-way criterion. However, for both CPM and MASPLAS, we see that the 2-way criterion is superior to the 1-way criterion. From Table 8.10, we see that 1-way criterion finds all the faults the earliest among the five criteria presented here, with 72 test cases for Book. For CPM and MASPLAS, 2-way criterion finds all the faults earliest with 346 and 58 test cases, respectively.

TABLE 8.10 Number of Tests Run before 100% Fault Detection

	Book	**CPM**	**MASPLAS**
1-Way	**72**	740	68
2-Way	83	**346**	**58**
Random	124	814	125
G-Best	19	480	27
G-Worst	124	890	164

8.5 TOOL: COMBINATORIAL-BASED PRIORITIZATION FOR USER-SESSION-BASED TESTING

The Combinatorial-based Prioritization for User-session-based Testing (CPUT) tool provides testers with the opportunity to convert user visits to test cases and further prioritize and reduce such test suites [172]. The tool has the following components: (1) an Apache module that formats POST/GET requests in the web log with the information needed to convert visits to test cases, (2) functionality to create a user-session-based test suite from a web log, and (3) functionality to prioritize or reduce test suites by combinatorial-based criteria. In the remainder of this section, we walk through each of these three components.

8.5.1 Apache Logging Module

The logger is implemented as a module for the Apache web server. It is written in C and records HTTP GET/POST requests and the associated data. To deploy the logger, the server administrator places the module in Apache's *modules* folder and enables the module through the configuration file. The log file that is generated by the logger contains a sequence of HTTP requests as received by the web server, including the IP address, time stamp, URL, associated parameter values, referrer URL, and cookie information.

8.5.2 Creating a User Session–Based Test Suite from Usage Logs Using CPUT

The test case creation engine of CPUT takes the log file generated by the logger and applies heuristics to create the test cases. Test cases are created based on the cookie, IP address, and pre-determined timeout interval [168]. User-session-based test cases are a sequence of HTTP requests and parameter name-value pairs stored in an XML format.

8.5.3 Prioritizing and Reducing Test Cases

CPUT can be used to prioritize and reduce test cases using one of several criteria. The combinatorial criteria discussed above are implemented in CPUT. After the user selects a criterion from a drop-down menu, the criterion is applied and the prioritized test order/reduced test suite is displayed to the user. The prioritized test order/reduced test suite can also be exported to a text file, which the user can use when executing the test cases. Figure 8.4 shows a screenshot of the CPUT tool.

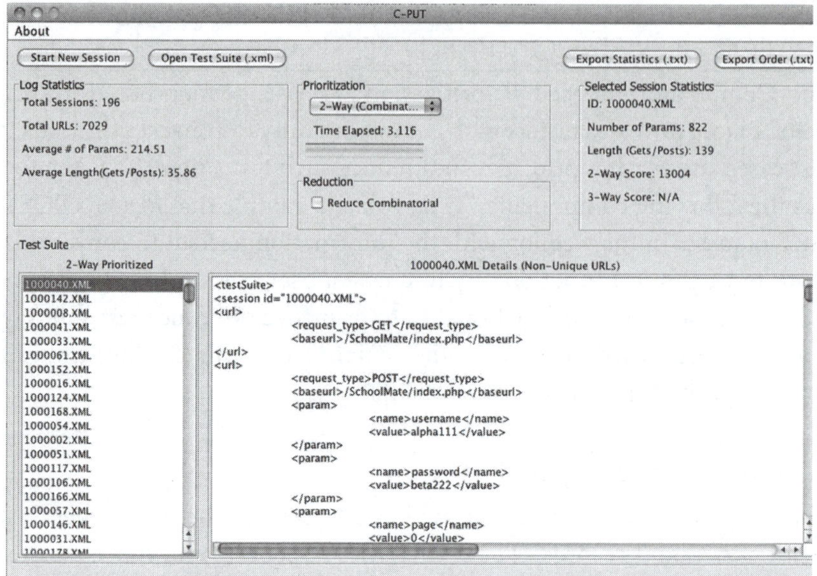

FIGURE 8.4 The XML test suite that was created from the web log.

In addition to the steps outlined here, CPUT also provides documentation and command line functionality to make scripting tests easy.

8.6 OTHER APPROACHES TO TEST SUITE PRIORITIZATION USING COMBINATORIAL INTERACTIONS

Bryce et al. [30] present a cost-cognizant algorithm for test suite prioritization based on interaction coverage between parameters. They consider the length of the test case in terms of the number of parameter values in the test as indicative of cost of the test case, since longer test cases take more time to execute than shorter test cases. The cost-based 2-way interaction coverage criterion proposed in this work selects a test case for the prioritized test order such that the test case maximizes the number of previously uncovered 2-way interactions divided by the number of parameters set to values in the test case. For two out of the three subject applications they evaluated in their experiments, they found that the cost-based 2-way interaction coverage criterion performed better than the 2-way interaction coverage criterion that did not consider cost. Different costs could also be used with the algorithm proposed in this work.

Bryce and Colbourn [26] create an algorithm to generate a test suite that provides combinatorial coverage in a prioritized manner. The algorithm allows users to specify weights to the parameter values. The weights are a value between 0 and 1 where the closer the weight is to 0, the less important it is and the closer to 1, the more important. The importance of a tuple is computed by multiplying the weights of the individual parameter values. The goal of the greedy algorithm then is to cover as much weight, that is, priority, as soon as possible.

Qu et al. [161] also prioritize regression test suites using combinatorial interactions between test cases. They use Bryce and Colburn's [26] algorithm to prioritize the test cases. The weights they use are based on code coverage. In the first weighting scheme, each value is assigned a weight that is equal to the number of occurrences of each value in the test suite ordered by cumulative code coverage, divided by the number of test cases in the test suite. The second weighting scheme follows the initial weighting as described above and then multiplies each value by a factor weight. The factor weight is determined as the maximum count of a value in the factor, m_i divided by the maximum of all individual factor maximums, m_i's. The final weighting scheme uses more of the code coverage information in the test suite and the first weighting scheme to assign weights to values. They also explore the use of weights based on specifications where weights are assigned based on intuition of which value will cause more code to be executed. In their experiments, they used two C programs, *flex* and *make*, with multiple versions. Through their experiments, they find that combinatorial-based prioritization is an effective regression testing method, however, the strength of interaction must be considered. In one of their applications, *flex*, they found that though combinatorial-based prioritization found faults sooner than prioritization by branch coverage, the technique missed a few faults. They also found no difference between the code coverage-based weighting scheme and the specification-based weighting scheme.

8.7 COST AND PRACTICAL CONSIDERATIONS

Efficiency improvements from prioritization can be especially useful for testing that requires human interaction.

Test suite prioritization is important when a test suite is larger and a tester desires to find more faults sooner in the testing process. Studies have shown that combinatorial-based prioritization criteria improve the rate of fault detection of the test suites, especially for event-driven software [28,171,161,29,30]. The CPUT tool [172] provides functionality for testers to quickly collect user-session-based test suites for web applications and prioritize by combinatorial coverage.

8.8 CHAPTER SUMMARY

Regression testing is an important part of the maintenance lifecycle of a software system. Selecting and ordering test cases is a challenge faced during regression testing. This chapter reviewed several combinatorial criteria for test suite prioritization. The criteria are based on the interactions between the parameters that are covered in a test case. The 1-way criterion focuses on covering 1-way interactions between parameters, and the 2-way criterion gives priority to test cases that cover pairs of parameters between windows. In studies on the effectiveness of combinatorial criteria on test case prioritization for event-driven software, researchers have found that the 1- and 2-way criteria perform better than random ordering, and are fairly close in performance to the theoretical best criterion, G-Best. We also discussed a tool, that is freely available for researchers and practitioners to use, that implements combinatorial prioritization for web application test cases.

REVIEW

Q: Define test suite prioritization.

A: Test suite prioritization is a regression testing problem, where test cases in a test suite are ordered using some criterion with the goal of achieving a performance goal.

Q: Given the following web application with 4 windows and corresponding parameter values, and the test suite with 3 test cases, list the 2-way inter-window interactions.

Window	Parameters
W1	p1, p2, p3, p4, p5
W2	p6, p7, p8, p9
W3	p10, p11, p12, p13, p14
W4	p15, p16, p17
W5	p18, p19

Test Case	Windows	Parameters
T1	W1-W2-W1-W3-W5	p1, p4, p5, p6, p7, p3, p10, p11, p19
T2	W2-W3-W4-W5	p6, p7, p10, p11, p13, p14, p16, p18
T3	W4-W5-W3	p15, p16, p18, p19, p10, p11

A: 2-way inter-window interactions are

Test Case	2-Way Interactions
T1	(p1, p6)(p1, p7)(p1, 10)(p1, p11)(p1, p19) (p4, p6)(p4, p7)(p4, 10)(p4, p11)(p4, p19) (p5, p6)(p5, p7)(p5, 10)(p5, p11)(p5, p19) (p6, p3)(p6, p10)(p6, p11)(p6, p19) (p7, p3)(p7, p10)(p7, p11)(p7, p19) (p3, p10)(p3, p11)(p3, p19) (p10, p19)(p11, p19)
T2	(p6, p10)(p6, p11)(p6, p13)(p6, p14)(p6, p16)(p6, p18) (p7, p10)(p7, p11)(p7, p13)(p7, p14)(p7, p16)(p7, p18) (p10, p16)(p10, p18) (p11, p16)(p11, p18) (p13, p16)(p13, p18) (p14, p16)(p14, p18) (p16, p18)
T3	(p15, p18)(p15, p19)(p15, p10)(p15, p11) (p16, p18)(p16, p19)(p16, p10)(p16, p11) (p18, p10)(p18, p11) (p19, p10)(p19, p11)

Q: Launch your own web application, set up the Apache module that comes with CPUT and have users visit your web application. Convert the web log to a user-session-based test suite using CPUT.

a. How many test cases are created?

b. How many 2-way inter-window interactions are in the test suite?

c. How many 2-way inter-window interactions are in the first test case?

d. How many 2-way inter-window interactions are in the last test case?

e. When you reduce the test suite by 2-way inter-window combinatorial coverage, how many tests are in the reduced test suite?

Q: Propose a future direction for combinatorial-based test suite prioritization.

A: There is no "one correct solution" to this question, but we suggest a few directions such as a new domain to apply the technique (i.e., embedded systems, unit testing, etc.) or a new way to measure combinatorial coverage (i.e., intra-window combinatorial coverage of parameters, a measure that considers constraints, etc.).

CHAPTER 9

Combinatorial Testing and Random Test Generation

For combinatorial testing to be most efficient and effective, we need an understanding of when a particular test development method is most appropriate. That is, what characteristics of a problem lead us to use one approach over another, and what are the tradeoffs with respect to cost and effectiveness? Some studies have reviewed the effectiveness of combinatorial and random approaches to testing, comparing the use of covering arrays with randomly generated tests, but have reached conflicting results [5,6,8,106,107,176]. Any single test containing values for n parameters, no matter how it is constructed, covers $C(n, 2)$ 2-way combinations (pairs), $C(n, 3)$ 3-way combinations, and so on. Naturally, as additional tests are added, more combinations are covered. A covering array packs combinations together closely, but as long a new test differs substantially from previously produced tests, additional combinations will be covered. Generating values randomly naturally leads to differences between tests, resulting in good combinatorial coverage for certain classes of problems. This chapter discusses the use of covering arrays and random test generation. As we will see, there is an interesting connection between these two concepts.

9.1 COVERAGE OF RANDOM TESTS

By definition, a covering array covers all t-way combinations for the specified value of t at least once. If enough random tests are generated, they will eventually also cover all t-way combinations. One key question is *how many random tests are needed to cover all t-way combinations*? In general, as the number of parameters increases, the probability that a random test set covers all t-way combinations increases as well, so in the limit with thousands of parameters, these two methods may begin to converge to the same number of tests. It has been shown [6] that where there is a large number of parameters (i.e., 1000s) and parameter values, and no constraints among parameters or parameter values, the number of tests required for t-way coverage (for arbitrary t) is approximately the same for covering arrays and randomly generated tests. This is an encouraging result, because of the difficulty of generating large covering arrays. We can produce thousands of random tests in seconds, but existing covering array algorithms cannot produce arrays for such large problems in a practical amount of time. If t-way coverage is needed for such problems, then random tests can be generated with a known probability of producing a full covering array. For N randomly generated tests containing k parameters with v_i values each, there is a probability P_t of detecting at least one t-way fault ($k > t$) [6]:

$$P_t \geq 1 - \left(1 - \frac{1}{\prod_{i=1}^{t} v_i}\right)^N \tag{9.1}$$

For the more common case where there are multiple faults, we need to also consider the ways in which combinations of faults can be discovered, leading to a probability $P_{t,z}$ to detect z different faults of [6]:

$$P_{t,z} \geq \sum_{j=0,z} (-1)^j \binom{z}{j} \left(1 - \frac{1}{\prod_{i=1}^{t} v_i}\right)^N \tag{9.2}$$

A large number of random tests can provide a high level of combinatorial coverage.

These probabilities converge to $\lim_{k\to\infty} P_t = 1$ and $\lim_{k\to\infty} P_{t,z} = 1$, for k parameters. For very large N, a randomly generated test set almost provides full t-way coverage. Although full coverage is not guaranteed because values are generated randomly, with enough tests we can obtain a very high probability of error detection. After all, our goal is not to use covering arrays, but to cover all t-way combinations, for the appropriate level of t. It does not matter how tests providing the necessary coverage are generated. As mentioned, an important caveat to this probability calculation is that it does not hold when constraints are involved, as they often are in practical testing problems. We can still generate tests randomly with constraints, but cannot rely on this calculation to estimate how many tests to produce.

For smaller test problems involving 10s of parameters, covering array algorithms are entirely practical and can cover all t-way combinations in a fraction of the number of tests required by random generation. Table 9.1 gives the percentage of t-way combinations covered by a randomly generated test set of the same size as a t-way covering array, for various combinations of k = number of variables and v = number of values per variable.

TABLE 9.1 Percent of t-Way Combinations Covered by Equal Number of Random Tests

Vars	Values/ Variable	ACTS 2-Way Tests	Random 2-Way Coverage (%)	ACTS 3-Way Tests	Random 3-Way Coverage (%)	ACTS 4-Way Tests	Random 4-Way Coverage (%)
10	2	10	89.28	20	92.18	42	92.97
10	4	30	86.38	151	89.90	657	92.89
10	6	66	84.03	532	91.82	3843	94.86
10	8	117	83.37	1214	90.93	12,010	94.69
10	10	172	82.21	2367	90.71	29,231	94.60
15	2	10	96.15	24	97.08	58	98.36
15	4	33	89.42	179	93.75	940	97.49
15	6	77	89.03	663	95.49	5243	98.26
15	8	125	85.27	1551	95.21	16,554	98.25
15	10	199	86.75	3000	94.96	40,233	98.21
20	2	12	97.22	27	97.08	66	98.41
20	4	37	90.07	209	96.40	1126	98.79
20	6	86	91.37	757	97.07	6291	99.21
20	8	142	89.16	1785	96.92	19,882	99.22
20	10	215	88.77	3463	96.85	48,374	99.20
25	2	12	96.54	30	98.26	74	99.18
25	4	39	91.67	233	97.49	1320	99.43
25	6	89	92.68	839	97.94	7126	99.59
25	8	148	90.46	1971	97.93	22,529	99.59
25	10	229	89.80	3823	97.82	54,856	99.58

Note that the coverage could vary with different realizations of randomly generated test sets. That is, a different random number generator, or even multiple runs of the same generator, may produce slightly different coverage (perhaps a few tests out of thousands, depending on the problem). Figures 9.1 through 9.6 summarize the coverage for arrays with variables of 2–10 values. As seen in the figures, the coverage provided by a random test suite versus a covering array of the same size varies considerably with different configurations.

Now consider the size of a random test set required to provide 100% combination coverage. With the most efficient covering array algorithms, the difficulty of finding tests with high coverage increases as tests are generated.

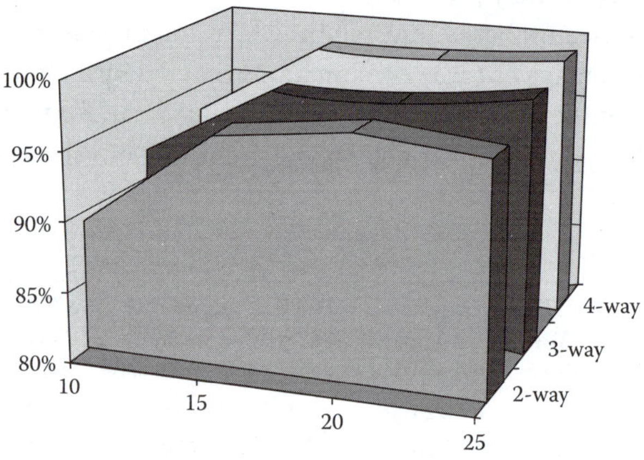

FIGURE 9.1 Percent coverage of t-way combinations for $v = 2$.

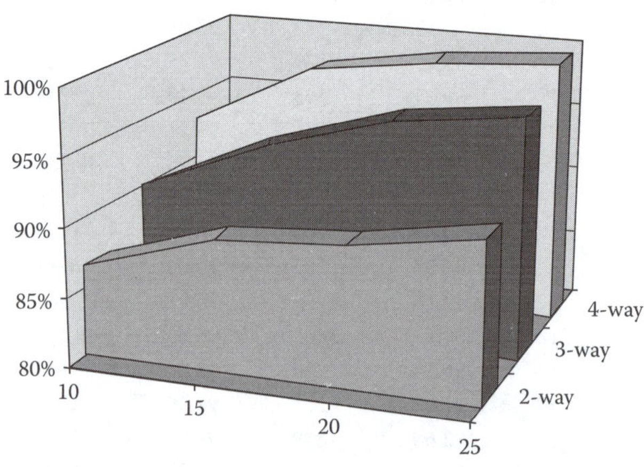

FIGURE 9.2 Percent coverage of t-way combinations for $v = 4$.

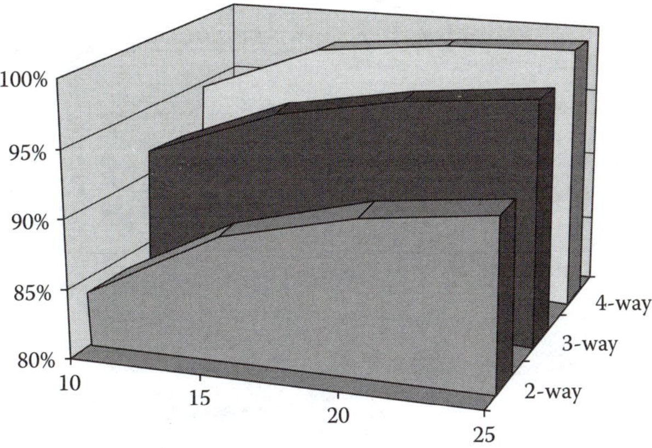

FIGURE 9.3 Percent coverage of t-way combinations for $v = 6$.

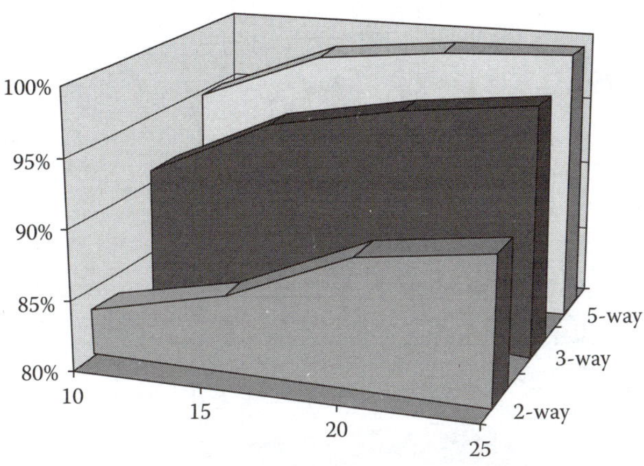

FIGURE 9.4 Percent coverage of t-way combinations for $v = 8$.

Thus, even if a randomly generated test set provides better than 99% of the coverage of an equal-sized covering array, it should not be concluded that only a few more tests are needed for the random set to provide 100% coverage. Table 9.2 gives the sizes of randomly generated test sets required for 100% combinatorial coverage at various configurations, and the ratio of these sizes to covering arrays computed with ACTS. Although there is considerable variation among configurations, note (Table 9.3) that the ratio of random to covering array size for 100% coverage exceeds 3 in most cases, with average ratios of 3.9, 3.8, and 3.2 at $t = 2$, 3, and 4, respectively. Thus, combinatorial testing retains a significant advantage over random testing for problems of this size if the goal is 100% combination coverage for a given value of t.

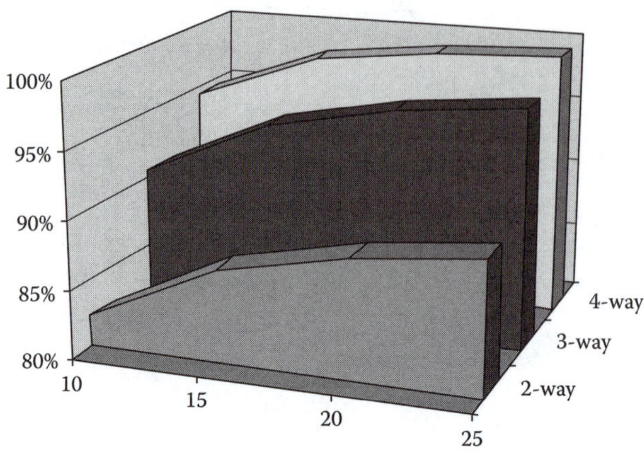

FIGURE 9.5 Percent coverage of t-way combinations for $v = 10$.

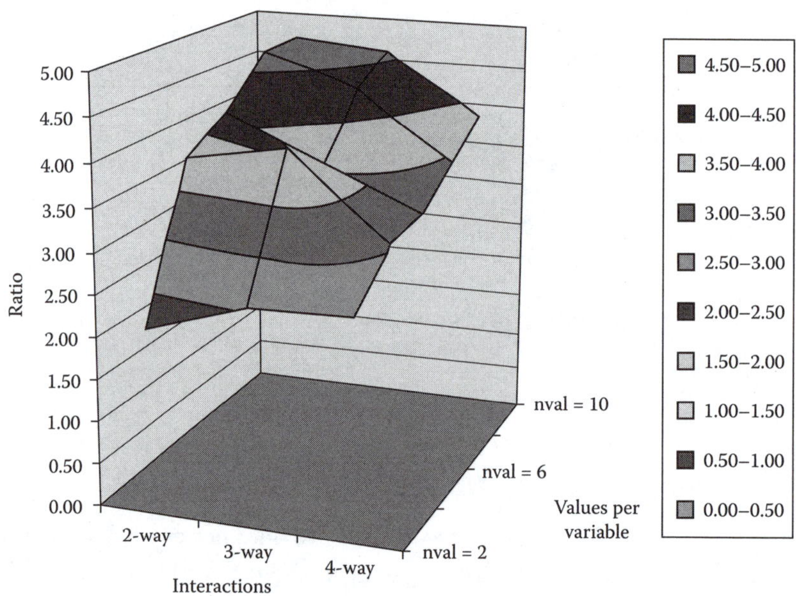

FIGURE 9.6 **(See color insert.)** Average ratio of random/ACTS for covering arrays by values per variable and interaction strength.

9.2 ADAPTIVE RANDOM TESTING

A recently developed testing strategy that may work quite well with combinatorial methods is called adaptive random testing (ART) [37–39]. The ART strategy seeks to deal with the problem that faults tend to cluster together [2,16,36], by choosing tests one at a time such that each newly

TABLE 9.2 Size of Random Test Set Required for 100% t-Way Combination Coverage

		2-Way Tests			3-Way Tests			4-Way Tests		
Vars	**Values**	**ACTS Tests**	**Random Tests**	**Ratio**	**ACTS Tests**	**Random Tests**	**Ratio**	**ACTS Tests**	**Random Tests**	**Ratio**
10	2	10	18	1.80	20	61	3.05	42	150	3.57
10	4	30	145	4.83	151	914	6.05	657	2256	3.43
10	6	66	383	5.80	532	1984	3.73	3843	13,356	3.48
10	8	117	499	4.26	1214	5419	4.46	12,010	52,744	4.39
10	10	172	808	4.70	2367	11,690	4.94	29,231	137,590	4.71
15	2	10	20	2.00	24	52	2.17	58	130	2.24
15	4	33	121	3.67	179	672	3.75	940	2568	2.73
15	6	77	294	3.82	663	2515	3.79	5243	17,070	3.26
15	8	125	551	4.41	1551	6770	4.36	16554	60,568	3.66
15	10	199	940	4.72	3000	15,234	5.08	40,233	159,870	3.97
20	2	12	23	1.92	27	70	2.59	66	140	2.12
20	4	37	140	3.78	209	623	2.98	1126	3768	3.35
20	6	86	288	3.35	757	2563	3.39	6291	18,798	2.99
20	8	142	630	4.44	1785	8450	4.73	19,882	59,592	3.00
20	10	215	1028	4.78	3463	14,001	4.04	48,374	157,390	3.25
25	2	12	34	2.83	30	70	2.33	74	174	2.35
25	4	39	120	3.08	233	790	3.39	1320	3520	2.67
25	6	89	327	3.67	839	2890	3.44	7126	19,632	2.75
25	8	148	845	5.71	1971	7402	3.76	22,529	61,184	2.72
25	10	229	1031	4.50	3823	16,512	4.32	54,856	191,910	3.50
Ratio average:				3.90			3.82			3.21

TABLE 9.3 Average Ratio of Random/ACTS for Covering Arrays by Values per Variable, Variables = 10, 15, 20, 25

Values per Variable	**Ratio, 2-Way**	**Ratio, 3-Way**	**Ratio, 4-Way**
2	2.14	2.54	2.57
4	3.84	4.04	3.04
6	4.16	3.59	3.12
8	4.70	4.33	3.44
10	4.68	4.59	3.86

chosen test is as "different" as possible from previous tests. The difference, or distance, metric is chosen based on problem characteristics. The basic ART algorithm is shown in Figure 9.7.

ART generates a set of random tests, determines the best test, that is, with the greatest distance from the existing test set T, then adds that test to T, continuing until some stopping criterion is fulfilled. If the distance

```
T= {} /* T is the set of previously executed test
   cases */
randomly generate a new input t for adding to T
test the program using t as a test case
add t to T
while (stopping criteria not reached) {
    D=0
    randomly generate next k candidates c1,c2, ... , ck
    for each candidate ci {
        calculate the minimum distance di from T
        if di>D {D=di; t=ci }
    }
    add t to T
    test the program using t as a test case
} // end while
```

FIGURE 9.7 Adaptive random testing algorithm.

metric is based on the number of previously uncovered t-way combinations that are covered in the candidate tests, then this algorithm is essentially a greedy algorithm [141] for computing a covering array one test at a time. The distance measures for this approach were originally developed for numeric processing. Many application domains, however, must deal with enumerated values with relatively little complex calculation. In these cases, distance measures tailored to covering arrays can help in choosing test order, that is, in prioritizing tests. Chapter 7 explains the use of prioritization methods.

9.3 TRADEOFFS: COVERING ARRAYS AND RANDOM GENERATION

The comparisons between random tests and covering arrays for combinatorial testing suggest a number of conclusions for many practical testing problems [106]:

- *For binary variables ($v = 2$), random tests compare reasonably well with covering arrays* (96–99% coverage) for all three values (2, 3, and 4) of t for 15 or more variables. Thus, random testing for a SUT with all or mostly binary variables may compare favorably with covering arrays.
- Combination coverage provided by random generation of the equivalent number of pairwise tests at ($t = 2$) decreases as the number of values per variable increases, and the coverage compared with pairwise testing is significantly <100%.

- *For 4-way interactions, coverage provided by random test generation increases with the number of variables.* Using a covering array for a module with approximately 10 variables should be significantly more effective than random testing, while the difference between the two methods should be less for modules with 20 or more variables.

- For 100% combination coverage, the efficiency advantage of covering arrays varies directly with the number of values per variable and inversely with the interaction strength t. Figure 9.6 illustrates how these factors (interaction strength t and values per variable v) combine: the ratio of random/covering array coverage is highest for 10 variables with $t = 2$, but declines for other pairings of t and v. To obtain 100% combination coverage, random testing is significantly less efficient, requiring two to nearly five times as many tests as a covering array generated by ACTS. Thus, if 100% combination coverage is desired, using covering arrays is less expensive than random test generation.

- For very large sets of parameters with no constraints, random test generation can produce a set of tests that cover all t-way combinations that is not significantly larger than the corresponding covering array. Generating the tests randomly will be much faster, and for very large problems covering array generation with existing tools is likely to be difficult.

An important practical consideration in comparing combinatorial with random testing is the efficiency of the covering array generator. Algorithms have a very wide range in the size of covering arrays they produce. In some cases, the better algorithms produce arrays that are 50% smaller than other algorithms. However, there is no uniformly "best" algorithm. Other algorithms may produce smaller or larger combinatorial test suites, so the comparable random test suite will vary in the number of combinations covered. Thus, random testing may fare better in comparison with combinatorial tests produced by one of the less-efficient algorithms.

A less optimal (by size) randomly generated array may provide better failure detection because it may include more interaction combinations of more than t variables.

However, there is a less obvious but important tradeoff regarding covering array size. An algorithm that produces a very compact array, that is, with few tests, for t-way combinations may include fewer $(t + 1)$-way combinations because there are fewer tests. Tables 9.4 and 9.5 illustrate this phenomenon for an example. Table 9.4 shows the percentage of $t + 1$ up to $t + 3$ combination coverage provided by the ACTS tests and in Table 9.5 the equivalent number of random tests. Although ACTS pairwise tests provide better 3-way coverage than the random tests, at other interaction strengths and values of t, the random tests are roughly the same or slightly better in combination coverage than ACTS. In one study [8], pairwise combinatorial tests detected slightly fewer events than the equivalent number of random tests. One possible explanation may be that the superior 4-way and 5-way coverage of the random tests allowed detection of more events. Almost paradoxically, an algorithm that produces a larger, sub-optimal covering array may provide better failure detection because the larger array is statistically more likely to include $t + 1$, $t + 2$, and higher degree interaction tests as a by-product of the test generation. Again, however, the less optimal covering array is likely to more closely resemble the random test suite in failure detection.

Note also that the number of failures in the SUT can affect the degree to which random testing approaches combinatorial testing effectiveness. For example, suppose the random test set covers 99% of combinations for 4-way interactions, and the SUT contains only one 4-way interaction failure. Then, there is a 99% probability that the random tests will contain

TABLE 9.4 Higher Interaction Coverage of t-Way Tests

t	3-Way Coverage	4-Way Coverage	5-Way Coverage
2	.758	.429	.217
3		.924	.709
4			.974

TABLE 9.5 Higher Interaction Coverage of Random Tests

t	3-Way Coverage	4-Way Coverage	5-Way Coverage
2	.735	.499	.306
3		.917	.767
4			.974

the 4-way interaction that triggers this failure. However, if the SUT contains m independent failures, then the probability that combinations for all m failures are included in the random test set is 0.99^m. Hence, with multiple failures, random testing may be significantly less effective, as its probability of detecting all failures will be c^m, for c = percent coverage and m = number of failures.

9.4 COST AND PRACTICAL CONSIDERATIONS

The relationship between covering arrays and randomly generated tests presents some interesting issues. Generating covering arrays for combinatorial tests is complex, but generating tests randomly is trivial. Thus, for large problems, we can compare the cost and time of generating a covering array versus producing tests randomly, measuring their coverage (Chapter 7), then adding tests as needed to provide full combinatorial coverage. Notice the last column of Table 9.1. For 4-way tests, once the number of parameters exceeds roughly 20, random generation will cover 99% or more of 4-way combinations. If a problem requires tests for 100 parameters, for example, covering array generators may require hours or days, or may simply be unable to handle that many parameters, but random tests could be generated quickly and easily. The test generation time for these two approaches is one factor among many that must be considered in test planning. Analyzing test parameters (Chapters 3 through 6), oracle development (Chapters 11 and 12), and other essential tasks such as test execution and managing test runs will generally be much more expensive than generating tests, regardless of the test generation method used.

The analyses reviewed in this chapter suggest a preference for tests based on covering arrays if the cost of applying the two test approaches is approximately the same. Most of the cost of testing goes into test planning, test oracle development, running and reporting tests, and the generation of test data—either randomly or with covering array tools—can be fully automated and run in parallel with other tasks. This preference may be particularly relevant if the SUT is likely to contain multiple failures (as is usually the case). Single failures that depend on the interaction of two or more variables have a high likelihood of being detected by random tests, because the random test set may cover a high percentage of all t-way combinations. But the probability of detecting multiple failures declines rapidly as c^m, for c = percent coverage and m = number of independent failures. Unfortunately, many testing problems are too large (too many parameters) to be handled entirely using covering arrays, so random test generation may be used to achieve the combinatorial coverage desired.

9.5 CHAPTER SUMMARY

Covering array algorithms are generally more efficient than random test generation if the goal is 100% combination coverage. Table 9.6 summarizes the test set size comparison for a variety of problem configurations. The difference is especially striking for binary parameters, where the ACTS covering array generator produces $(t + 1)$-way coverage with roughly the same number of tests required by random generation for t-way coverage. Table 9.2 provides additional information.

Existing research has shown either no difference (for some problems) or higher failure detection effectiveness (for most problems) for covering arrays as compared with randomly generated tests. Analyzing random test sets suggests a number of reasons for this result. In particular, a highly optimized t-way covering array may include fewer $t + 1$, $t + 2$, and higher degree interaction tests than an equivalent sized random test set. Similarly, a covering array algorithm that produces a larger, sub-optimal array may provide better failure detection because the larger array is statistically more likely to include $t + 1$, $t + 2$, and higher degree interaction tests as a by-product of the test generation. In some applications, it may make sense to combine aspects of both approaches.

TABLE 9.6 ACTS and Random Test Set Sizes for 100% t-Way Combination Coverage

		$t = 2$		$t = 3$		$t = 4$	
n	v	ACTS	Random	ACTS	Random	ACTS	Random
10	2	10	18	20	61	42	150
10	4	30	145	151	914	657	2256
10	8	117	499	1214	5419	12,010	52,744
15	2	10	20	24	52	58	130
15	4	33	121	179	672	940	2568
15	8	125	551	1551	6770	16,554	60,568
20	2	12	23	27	70	66	140
20	4	37	140	209	623	1126	3768
20	8	142	630	1785	8450	19,882	59,592
25	2	12	34	30	70	74	174
25	4	39	120	233	790	1320	3520
25	8	148	845	1971	7402	22,529	61,184

REVIEW

Q: The U.S. Federal Aviation Administration requires that Level A (life critical) software have a failure probability of no more than 10^{-9}. If combinatorial testing were sufficient for such a level of assurance (it is not, this is just an exercise!), how many tests would be required for a system with all binary parameters to detect at least one 2-way interaction fault with a probability of $1–10^{-9}$? Use Equation 9.1.

A: 73 tests should do it.

Q: If the system has 2000 parameters, how large a covering array would be needed to detect 100% of the 2-way faults? Use the Table: http://math.nist.gov/coveringarrays/ipof/Tables/Table.2.2.html

A: 21 tests. This example illustrates that when there are few values per parameter, even a large number of parameters can be handled with combinatorial testing for low values of *t*. Even with 6 values per parameter, only 195 tests are needed for 2000 parameters.

Q: The Interaction Rule states that the number of failures decreases rapidly with increasing levels of *t*, but research shows that 5-way and 6-way faults happen in fielded systems. How many random tests are needed to assure the system with all binary parameters to FAA requirements if we want to catch 5-way faults?

A: 655 tests.

Q: If the system has 2000 parameters, can you compute a 5-way covering array for the preceding problem?

A: Not likely with current tools. This example illustrates a case where random testing would be a good alternative. But remember that Equation 9.1 gives the probability of catching at least one fault, while the true number is unknown (that's why we're testing!) Thus, we may want to generate tests randomly, then measure coverage and supplement the test set for full 5-way coverage.

Q: What if the problems described above involve a variety of constraints among parameter values. Can you still apply Equation 9.1 to compute the probability of detecting a fault?

A: Sadly, no.

CHAPTER 10

Sequence-Covering Arrays

In testing event-driven software, the critical condition for triggering failures often is whether or not a particular event has occurred prior to a second one, establishing a particular state that must be reached before a given failure can be triggered. For example, a failure might occur when connecting device *A* only if device *B* is already connected, or only if devices *B* and *C* were both already connected. Events may be repeatable in some systems, but this is not always the case. In the testing problem that motivated this work, the critical issue was the sequence of connecting a large number of peripherals, so it was physically impossible to connect an already-connected device (without unplugging, which would be a separate event). As a different example, a memory management function may fail on an attempt to allocate memory if it failed to properly release memory at some prior time. Another common class of problems of this type occurs with graphical user interfaces (GUIs) that use callbacks. User actions may trigger the creation or release of resources, or the enabling or disabling of GUI controls. But the user may invoke these callbacks in any order, and errors may result if a prior callback left the system in an unexpected state.

10.1 SEQUENCE-COVERING ARRAY DEFINITION

In many systems, the order of inputs is important.

TABLE 10.1 System Events

Event	Description
a	Connect air flow meter
b	Connect pressure gauge
c	Connect satellite link
d	Connect pressure readout
e	Engage drive motor
f	Engage steering control

For this problem, we can define a *sequence-covering array* [11,67,108–111], which is a set of tests that ensure that all *t-way sequences* of events have been tested. The *t* events in the sequence may be interleaved with others, but all permutations will be tested. For example, we may have a component of a factory automation system that uses certain devices interacting with a control program. We want to test the events defined in Table 10.1.

There are 6! = 720 possible sequences for these six events, and the system should respond correctly and safely regardless of the order in which they occur. Operators may be instructed to use a particular order, but mistakes are inevitable, and should not result in injury to users or compromise the enterprise. Because setup, connections, and operation of this component are manual, each test can take a considerable amount of time. It is not uncommon for system-level tests such as this to take hours to execute, monitor, and complete. We want to test this system as thoroughly as possible, but time and budget constraints do not allow for testing all possible sequences, so we will test all three-event sequences.

With six events, *a, b, c, d, e,* and *f,* one subset of three is {*b, d, e*}, which can be arranged in six permutations: [*b d e*], [*b e d*], [*d b e*], [*d e b*], [*e b d*], [*e d b*]. A test that covers the permutation [*d b e*] is [*a d c f b e*]; another is [*a d c b e f*]. A larger example system may have 10 devices to connect, in which case the number of permutations is 10!, or 3,628,800 tests for exhaustive testing. In that case, a three-way sequence-covering array with 14 tests covering all $10 \cdot 9 \cdot 8 = 720$ three-way sequences is a dramatic improvement, as is 72 tests for all four-way sequences (see Table 10.3 later in the chapter).

Definition

A sequence-covering array, SCA(N, S, t) is an $N \times S$ matrix, where entries are from a finite set S of s symbols, such that every t-way permutation of symbols from S occurs in at least one row; the t symbols in the permutation

TABLE 10.2 Tests for Four Events

Test				
1	*a*	*d*	*b*	*c*
2	*b*	*a*	*c*	*d*
3	*b*	*d*	*c*	*a*
4	*c*	*a*	*b*	*d*
5	*c*	*d*	*b*	*a*
6	*d*	*a*	*c*	*b*

are not required to be adjacent. That is, for every *t*-way arrangement of symbols x_1, x_2, …, x_t, the regular expression $.^*x_1.^*x_2 \ldots .^*x_t.^*$ matches at least one row in the array. Sequence-covering arrays, as the name implies, are analogous to standard covering arrays (see Section 1.2), which include at least one of every *t*-way combination of any *n* variables, where $t < n$. A variety of algorithms are available for constructing covering arrays, but these are not usable for generating *t*-way sequences because they are designed to cover combinations in any order.

Example

Consider the problem of testing four events, *a, b, c,* and *d*. For convenience, a *t*-way permutation of symbols is referred to as a *t-way sequence*. There are $4! = 24$ possible permutations of these four events, but we can test all three-way sequences of these events with only six tests (see Table 10.2).

10.2 SIZE AND CONSTRUCTION OF SEQUENCE-COVERING ARRAYS

Sequence-covering arrays can be constructed with a variety of methods. A two-way sequence-covering array can be constructed simply by listing the events in some order for one test and in reverse order for the second test:

1	*a*	*b*	*c*	*d*
2	*d*	*c*	*b*	*a*

To see that this procedure generates tests that cover all two-way sequences, note that for two-way sequence coverage, every pair of variables *x* and *y*, *x*…*y* and *y*…*x*, must both be in some test (where *a*…*b* means that *a* is eventually followed by *b*). All variables are included in each test; therefore, any sequence *x*…*y* must be in either test 1 or test 2 and its reverse *y*…*x* in the other test. Thus, only two tests are needed to cover all two-way sequences, regardless of the number of events to be included in the tests. This can be an effective way of doing initial tests on a GUI with multiple buttons, text input boxes, selection lists, and other features. Invoking each of the features on screen in some order and then reversing the order may uncover problems in memory management or initialization (often as a result of developers' assumptions about the order in which the user will interact with the system).

TABLE 10.3 Number of Tests for Combinatorial Three-Way and Four-Way Sequences

Events	Three-Sequence Tests	Four-Sequence Tests
5	8	29
6	10	38
7	12	50
8	12	56
9	14	68
10	14	72
11	14	78
12	16	86
13	16	92
14	16	100
15	18	108
16	18	112
17	20	118
18	20	122
19	22	128
20	22	134
21	22	134
22	22	140
23	24	146
24	24	146
25	24	152
26	24	158
27	26	160
28	26	162
29	26	166
30	26	166
40	32	198
50	34	214
60	38	238
70	40	250
80	42	264

The number of tests required for t-way coverage of n events and the lower bound for a sequence-covering array grows logarithmically in n [109]. Therefore, a large number of events can be tested using a reasonable number of tests for most applications, as can be confirmed in Table 10.3. Greedy methods produce good results across a broad range of problem sizes. Construction methods for sequence-covering arrays also include answer-set programming [11,67]. Answer-set programming can generate more compact test sets than greedy methods, but this advantage may not hold for larger problem sizes.

10.2.1 Generalized t-Way Sequence Covering

For *t-way sequence* test generation, where $t > 2$, one method is to use a greedy algorithm (Figure 10.1) that generates a large number of tests, scores each by the number of previously uncovered sequences it covers, then chooses the highest scoring test. This simple approach produces surprisingly good results, in both test set size and execution time.

10.2.2 Algorithm Analysis

The complexity of the algorithm is dominated by the selection of a candidate test that covers the greatest number of previously uncovered sequences. An array of bits for each possible t-way sequence is used so that marking and testing the array for a particular sequence can be done in constant time for each of the t-way sequences. This selection process checks each of the $n \times (n-1) \times \cdots \times (n-t+1)$ possible t-way sequences to determine if the sequence has previously been covered or is newly covered by the candidate test. The check is done for each of the N candidate

```
Algorithm t-seq(int t, int n)
// t = interaction strength; n = # parameters, n > t;
 N = # candidate tests to generate
 initialize test set ts to be an empty set;
 initialize set chk of n×(n−1)×...×(n−t+1) bits to 0;
 while (all t-way sequences not marked in chk) {
   1. tc := set of N test candidates generated with random values of each of the n parameters
   2. test1 := test from set tc that covers the greatest number of sequences not marked as covered
        in chk;
   3. for each new sequence covered in test1, mark corresponding bit in set chk to 1;
   4. ts := ts U test1 ;
   5. if (symmetry && all t-way sequences not marked in chk) { test2 := reverse(test1);
        ts := ts U test2 ;
        for each new sequence cover in test2,
           mark corresponding bit in set chk to 1; }
 }
 return ts;
```

FIGURE 10.1 Algorithm t-sequence.

tests, with constant N, so the time complexity of the algorithm is $O(n^t)$. Storage required for the algorithm is also $O(n^t)$, because of the set *chk* for keeping track of which sequences have been covered at each step.

It is shown in Ref. [109] that the number of tests generated by a greedy algorithm grows logarithmically with n. At each step, a greedy algorithm that selects the test that covers the largest number of previously uncovered sequences will progress at a rate of at least $1/t!$ of the remaining sequences at each iteration. Thus, uncovered sequences are reduced as $U_{i+1} = U_i(1 - 1/t!)$, and after k iterations, remaining uncovered sequences will be $U_0(1 - 1/t!)^k$. Initially, $U_0 = n \times (n-1) \times \cdots \times (n-t+1)$. For small n, it may be possible to implement an optimal greedy algorithm that tests all $n!$ possible tests. For larger values of n, the algorithm may be reasonably close to finding an optimal next test, with sufficient candidates.

10.3 USING SEQUENCE-COVERING ARRAYS

Sequence-covering arrays have been incorporated into operational testing for a mission-critical system that uses multiple devices with inputs and outputs to a laptop computer. The test procedure has eight steps: boot system, open application, run scan, and connect peripherals P-1 through P-5. It is expected that for some sequences, the system will not function properly, thus the order of connecting peripherals is a critical aspect of testing. In addition, there are constraints on the sequence of events: cannot scan until the app is open; cannot open app until system is booted. There are 40,320 permutations of eight steps, but some are redundant (e.g., changing the order of peripherals connected before boot), and some are invalid (violates a constraint). Around 7000 are valid, and nonredundant, but this is far too many to test for a system that requires manual, physical connections of devices.

The system was tested using a seven-step sequence-covering array, incorporating the assumption that there is no need to examine strength-3 sequences that involve boot-up. The initial test configuration (Table 10.4) was drawn from the library of precomputed sequence tests. Some changes were made to the precomputed sequences based on unique requirements of the system test. If 6 = "Open App" and 5 = "Run Scan," then cases 1, 4, 6, 8, 10, and 12 are invalid, because the scan cannot be run before the application is started. This was handled by "swapping 0 and 1" when they are adjacent (1 and 4), out of order. For the other cases, several cases were generated from each that were valid mutations of the invalid case. A test was also embedded to see whether it mattered where each of three USB connections was placed. The last test case ensures at least strength 2 (sequence of length 2) for all

TABLE 10.4 Seven-Event Tests from Precomputed Test Library

Test 1	0	1	2	3	4	5	6
Test 2	6	5	4	3	2	1	0
Test 3	2	1	0	6	5	4	3
Test 4	3	4	5	6	0	1	2
Test 5	4	1	6	0	3	2	5
Test 6	5	2	3	0	6	1	4
Test 7	0	6	4	5	2	1	3
Test 8	3	1	2	5	4	6	0
Test 9	6	2	5	0	3	4	1
Test 10	1	4	3	0	5	2	6
Test 11	2	0	3	4	6	1	5
Test 12	5	1	6	4	3	0	2

peripheral connections and "Boot," that is, that each peripheral connection occurs prior to boot. The final test array is shown in Table 10.5.

10.4 COST AND PRACTICAL CONSIDERATIONS

As with other forms of combinatorial testing, some combinations may be either impossible or not exist on the system under test. For example, "receive message" must occur before "process message." One algorithm for sequence-covering arrays makes it possible to specify pairs *x*,*y*, where the sequence *x*...*y* is to be excluded from the generated covering array. Typically, this will lead to extra tests, but does not increase the test array significantly.

Sequence covering can be relatively inexpensive as a test technique. As noted previously, only two tests are needed to produce two-way covering, and the number of tests grows only as log n for n events for $t > 2$. Table 10.3 shows the number of tests for three-way and four-way sequences. Different algorithms may produce fewer or more tests than shown in Table 10.3.

10.5 CHAPTER SUMMARY

Sequence-covering arrays are a new application of combinatorial methods, developed to solve problems with interoperability testing. A sequence-covering array is a set of tests that ensures that all *t-way sequence*s of events have been tested. The t events in the sequence may be interleaved with others, but all permutations will be tested. All two-way sequences can be tested simply by listing the events to be tested in any order, then reversing the order to create a second test. Algorithms have been developed to create sequence-covering arrays for higher-strength interaction

TABLE 10.5 Final Sequence-Covering Array Used in Testing the System Set Up Example

Original Case	Case	Step 1	Step 2	Step 3	Step 4	Step 5	Step 6	Step 7	Step 8
1	1	Boot	P-1 (USB-RIGHT)	P-2 (USB-BACK)	P-3 (USB-LEFT)	P-4	P-5	Application	Scan
2	2	Boot	Application	Scan	P-5	P-4	P-3 (USB-RIGHT)	P-2 (USB-BACK)	P-1 (USB-LEFT)
3	3	Boot	P-3 (USB-RIGHT)	P-2 (USB-LEFT)	P-1 (USB-BACK)	Application	Scan	P-5	P-4
4	4	Boot	P-4	P-5	Application	Scan	P-1 (USB-RIGHT)	P-2 (USB-LEFT)	P-3 (USB-BACK)
5	5	Boot	P-5	P-2 (USB-RIGHT)	Application	P-1 (USB-BACK)	P-4	P-3 (USB-LEFT)	Scan
6A	6	Boot	Application	P-3 (USB-BACK)	P-4	P-1 (USB-LEFT)	Scan	P-2 (USB-RIGHT)	P-5
6B	7	Boot	Application	Scan	P-3 (USB-LEFT)	P-4	P-1 (USB-RIGHT)	P-2 (USB-BACK)	P-5
6C	8	Boot	P-3 (USB-RIGHT)	P-4	P-1 (USB-LEFT)	Application	Scan	P-2 (USB-BACK)	P-5
6D	9	Boot	P-3 (USB-RIGHT)	Application	P-4	Scan	P-1 (USB-BACK)	P-2 (USB-LEFT)	P-5
7	10	Boot	P-1 (USB-RIGHT)	Application	P-5	Scan	P-3 (USB-BACK)	P-2 (USB-LEFT)	P-4

8A	11	Boot	P-4	P-2 (USB-RIGHT)	P-3 (USB-LEFT)	Application	Scan	P-5	P-1 (USB-BACK)
8B	12	Boot	P-4	P-2 (USB-RIGHT)	P-3 (USB-BACK)	P-5	Application	Scan	P-1 (USB-LEFT)
9	13	Boot	Application	P-3 (USB-LEFT)	Scan	P-1 (USB-RIGHT)	P-4	P-5	P-2 (USB-BACK)
10A	14	Boot	P-2 (USB-BACK)	P-5	P-4	P-1 (USB-LEFT)	P-3 (USB-RIGHT)	Application	Scan
10B	15	Boot	P-2 (USB-LEFT)	P-5	P-4	P-1 (USB-BACK)	Application	Scan	P-3 (USB-RIGHT)
11	16	Boot	P-3 (USB-BACK)	P-1 (USB-RIGHT)	P-4	P-5	Application	P-2 (USB-LEFT)	Scan
12A	17	Boot	Application	Scan	P-2 (USB-RIGHT)	P-5	P-4	P-1 (USB-BACK)	P-3 (USB-LEFT)
12B	18	Boot	P-2 (USB-RIGHT)	Application	Scan	P-5	P-4	P-1 (USB-LEFT)	P-3 (USB-BACK)
NA	19	P-5	P-4	P-3 (USB-LEFT)	P-2 (USB-RIGHT)	P-1 (USB-BACK)	Boot	Application	Scan

levels. For a given interaction strength, the number of tests generated is proportional to the log of the number of events. As with other types of combinatorial testing, constraints may be important, since it is very common that certain events depend on others occurring first. The tools developed for this problem allow the user to specify constraints in the form of excluded sequences, which will not appear in the generated test array.

REVIEW

Q: Given 10 possible system events, how many possible sequences of the events can be arranged?

A: 10! = 3,628,800.

Q: How many tests are needed to cover all two-way sequences of 10 events? 20 events?

A: Only two tests are needed for two-way coverage of any number of events.

Q: At what rate does the number of tests increase as the number of events increases, for a given level of t?

A: The number of tests increases proportional to log n, for n events.

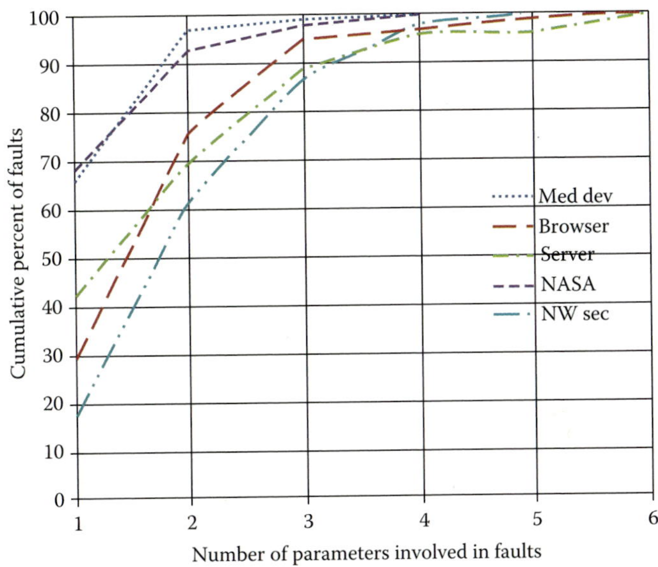

FIGURE 1.2 The Interaction Rule: Most failures are triggered by one or two parameters interacting, with progressively fewer by 3, 4, or more.

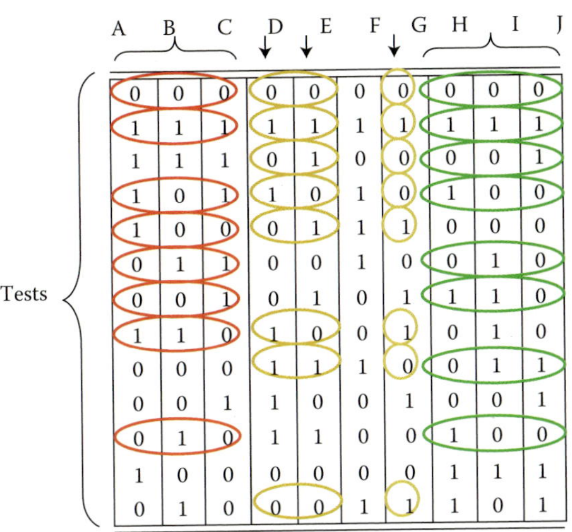

A	B	C	D	E	F	G	H	I	J
0	0	0	0	0	0	0	0	0	0
1	1	1	1	1	1	1	1	1	1
1	1	1	0	1	0	0	0	0	1
1	0	1	1	0	1	0	1	0	0
1	0	0	0	1	1	1	0	0	0
0	1	1	0	0	1	0	0	1	0
0	0	1	0	1	0	1	1	1	0
1	1	0	1	0	0	1	0	1	0
0	0	0	1	1	1	0	0	1	1
0	0	1	1	0	0	1	0	0	1
0	1	0	1	1	0	0	1	0	0
1	0	0	0	0	0	0	1	1	1
0	1	0	0	0	1	1	1	0	1

FIGURE 1.3 A 3-way covering array includes all 3-way combinations of values.

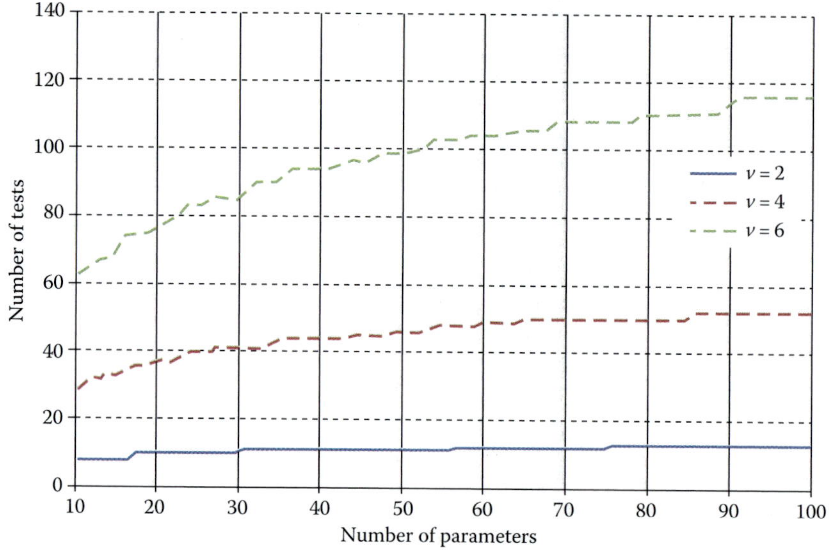

FIGURE 1.4 Number of tests, $t = 2$.

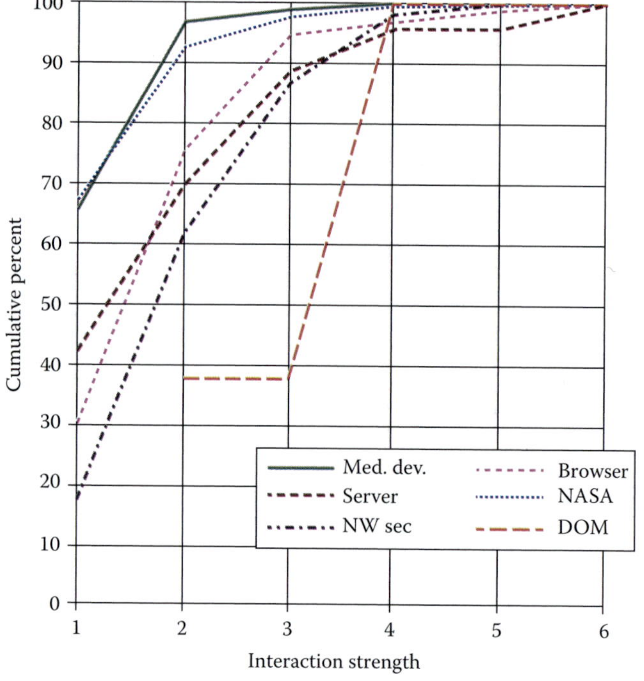

FIGURE 2.2 DOM compared with other applications.

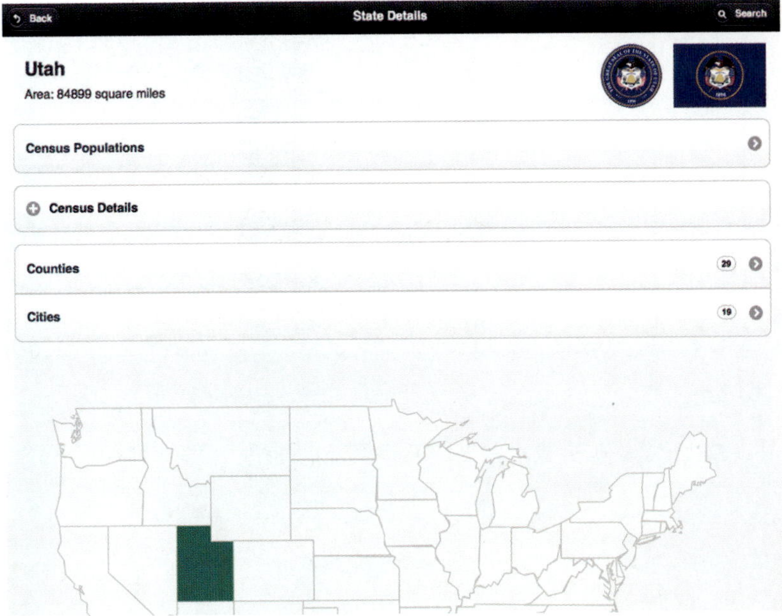

FIGURE 2.4 Screenshot of the sample RWA.

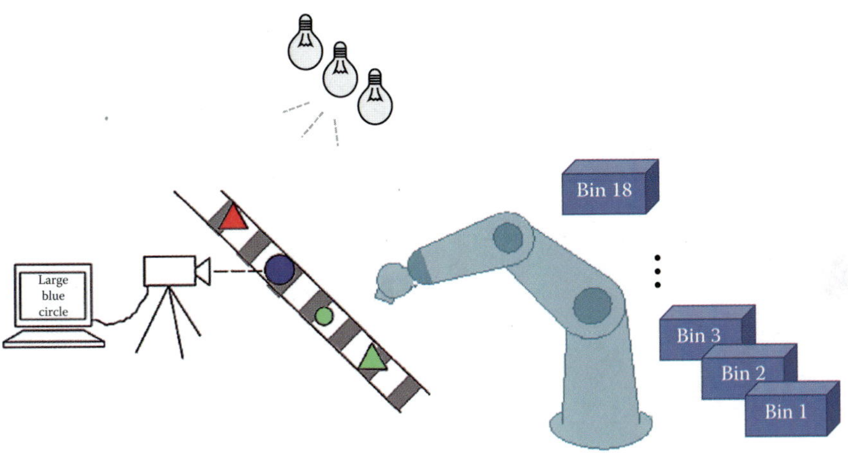

FIGURE 5.3 The flexible manufacturing system.

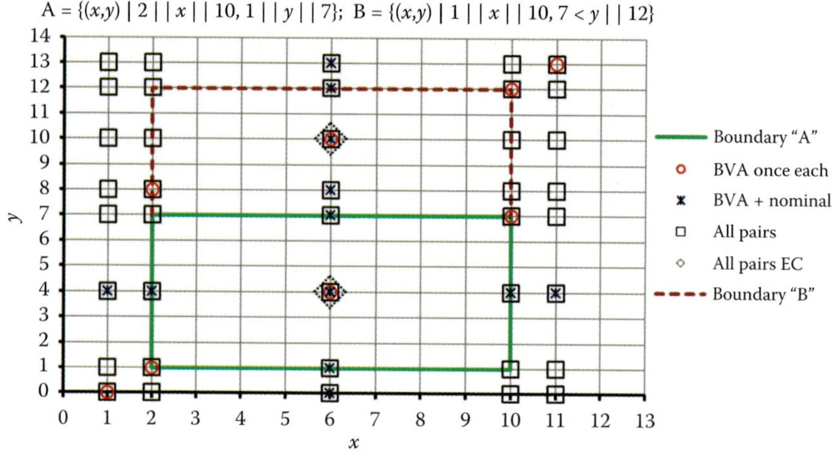

FIGURE 5.21 A two-dimensional convex region tested using different strategies.

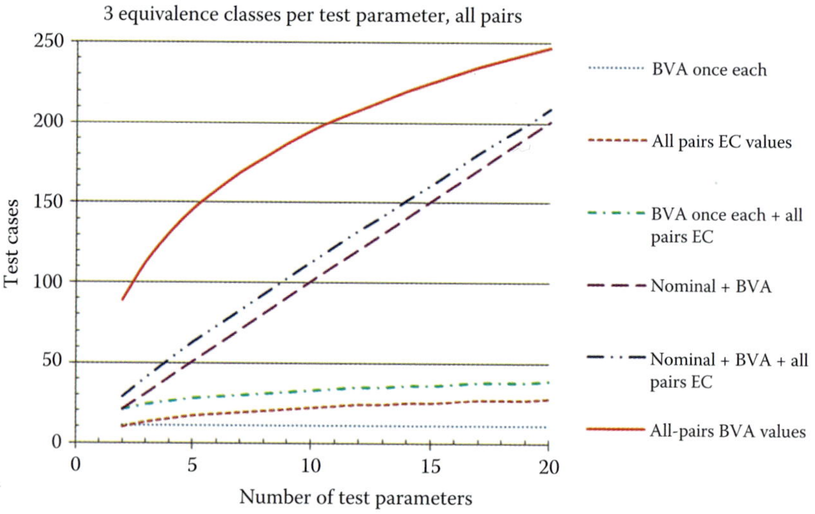

FIGURE 5.22 Number of test cases required by different testing strategies.

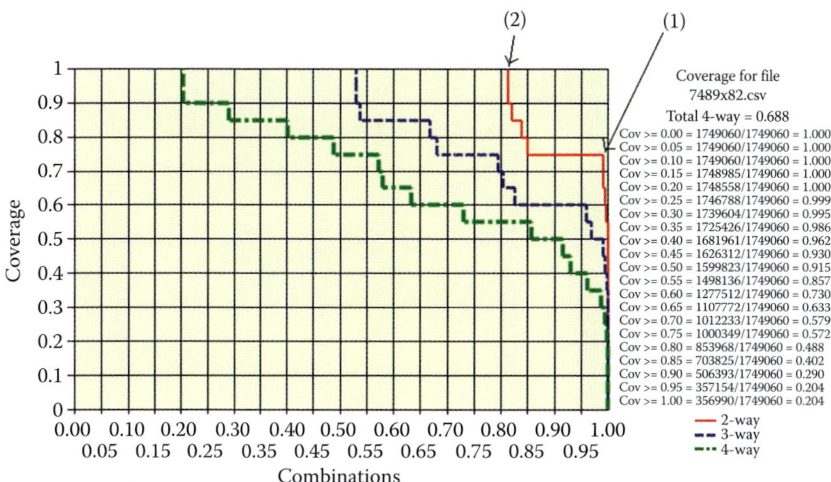

FIGURE 7.7 Configuration coverage for $1^3 2^{75} 4^2 6^2$ inputs. (1) Configurations covered to 75% level and (2) configurations covered to 100% level.

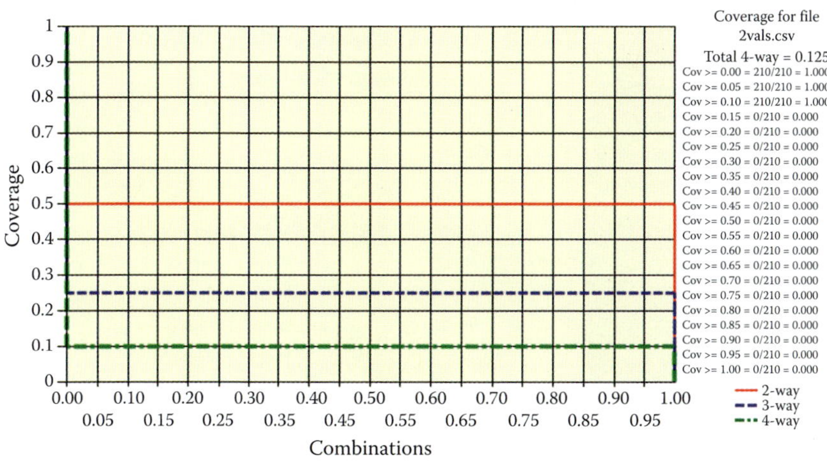

FIGURE 7.10 *t*-Way coverage for two tests with binary values.

FIGURE 8.1 Screenshot of the SchoolMate page for "Add New Attendance Record."

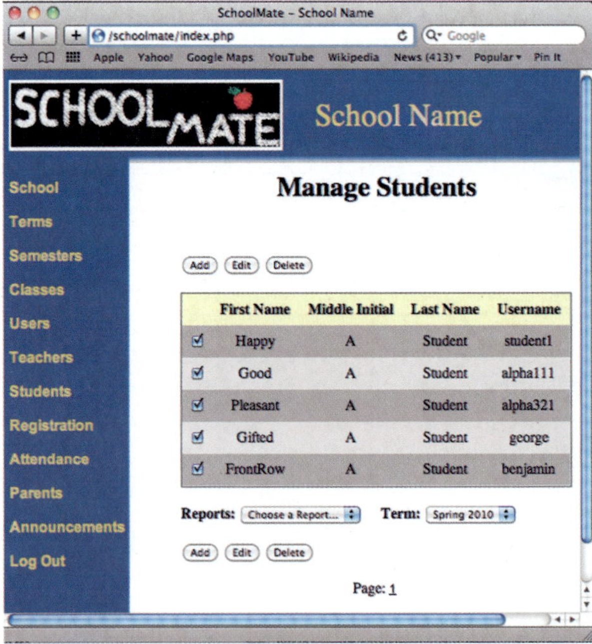

FIGURE 8.2 Screenshot of the SchoolMate page for "Manage Students."

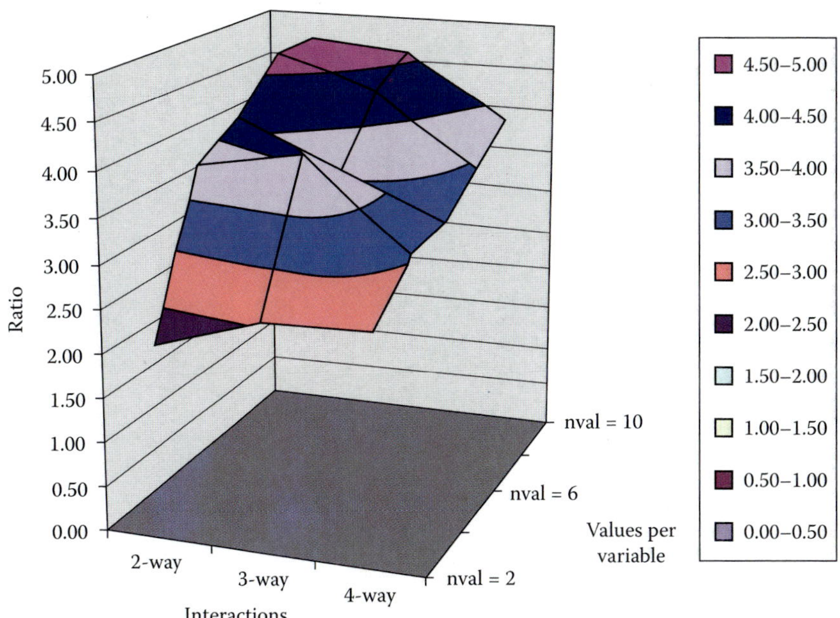

FIGURE 9.6 Average ratio of random/ACTS for covering arrays by values per variable and interaction strength.

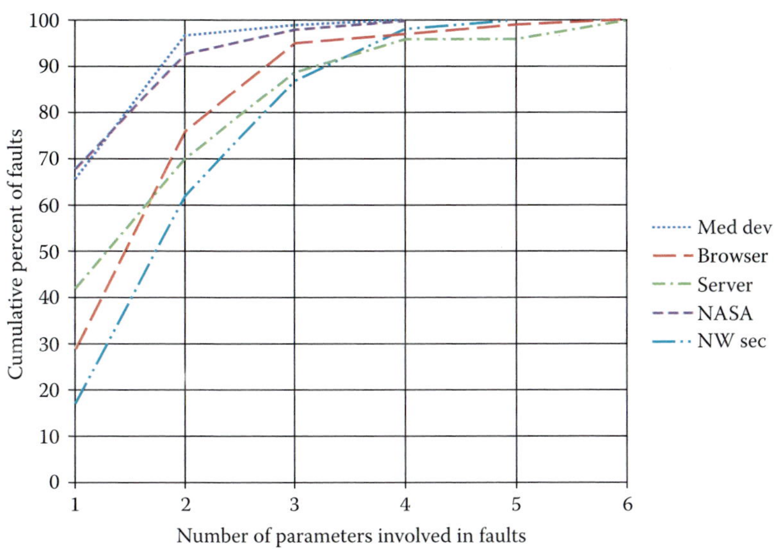

FIGURE B.1 Cumulative percentage of failures triggered by *t*-way interactions.

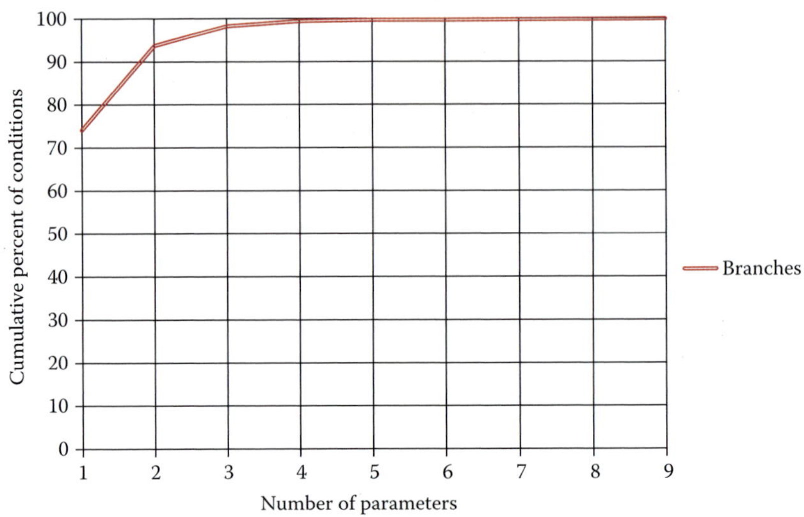

FIGURE B.2 Cumulative percentage of branches containing *n* variables.

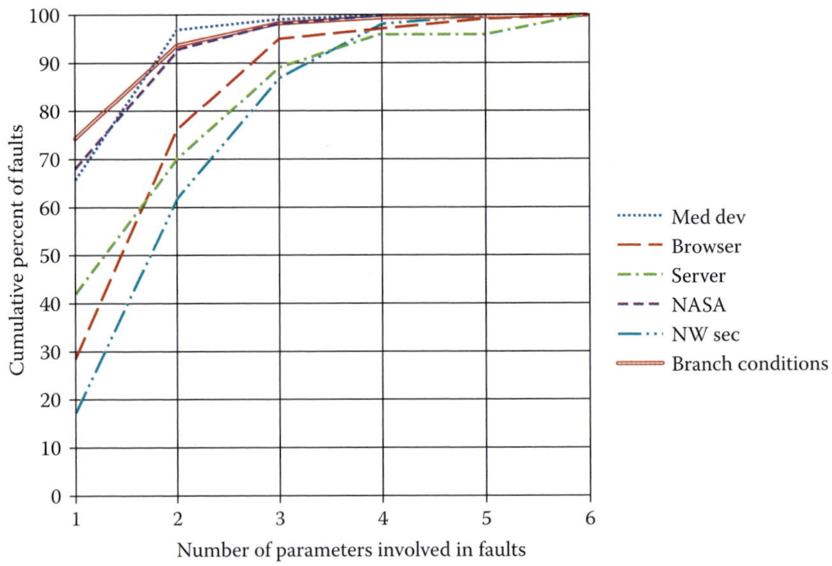

FIGURE B.3 Branch distribution (green) superimposed in Figure B.1.

CHAPTER 11

Assertion-Based Testing

BUILT-IN SELF-TEST IS A common feature for integrated circuits in which additional hardware and software functions allow checking for correct operation while the system is running, in contrast with the use of externally applied tests. A similar concept—embedded assertions—has been used for decades in software, and advances in programming languages and processor performance have made this method even more useful and practical. It can be especially effective with combinatorial testing. If the system is fully exercised with t-way tests, and assertions are thorough, we can have reasonable confidence that operation will be correct for up to t-way combinations of input values. In addition to standard programming language features, a variety of specialized tools have been developed to make this approach easier and more effective.

Many programming languages include an *assert* feature that allows the programmer to specify properties that are assumed true at a particular point in the program. For example, a function that includes a division in which a particular parameter x will be used as a divisor may require that this parameter may never be zero. This function may include the C statement `assert(x  ! = 0);` as the first statement executed. Note that the assertion is not the same as an input validity check that issues an error message if input is not acceptable. The assertion specifies conditions that must hold for the function to operate properly, in this case, a nonzero divisor. The distinction between assertions and input validation code is that assertions are intended to catch *programming mistakes*, while input validation detects errors in user or file/database input.

With self-testing through assertions, thousands of tests can often be run at very low cost, allowing high-strength interaction coverage.

With a sufficient number of assertions derived from a specification, the program can have a self-checking property [86,138,124]. The assertions can serve as a sort of embedded proof of important properties, such that if the assertions pass for all executions of the program, then the properties encoded in the assertions can be expected to hold. Then, if the assertions form a chain of logic that implies a formal statement of program properties, the program's correctness with respect to these properties can be proven. We can take advantage of this scheme in combinatorial testing by demonstrating that the assertions hold for all *t*-way combinations of inputs. While this is not the same as a correctness proof, it is an effective way of integrating formal methods for correctness with program testing, and an extensive body of research has developed this idea for practical use (for a survey, see Ref. [12]). Modern programming languages include support for including assertions that encode program properties, and tools such as the Java modeling language [115] have been designed to integrate assertions with testing. In many cases, using assertions to self-check important properties makes it practical to run thousands of tests in a fully automated fashion, so high-strength interactions of four-way and above can be done in reasonable time. Since important properties of the system are checked with every run, by executing the code with all *t*-way combinations of input, we can have high confidence that it works correctly.

11.1 BASIC ASSERTIONS FOR TESTING

To clarify this somewhat abstract discussion, we will analyze requirements for a small function that handles withdrawal processing for an automated teller machine (ATM). Graphical user interface code for the ATM will not be displayed, as this would vary considerably for different systems. The decision not to include GUI code in this example also illustrates a practical limitation of this type of testing: there are many potential sources of error in a software project, and testing may not deal with all of them at the same time. The GUI code may be analyzed separately, or a more complex verification with assertions may specify properties of the GUI calls, but in the end, some human involvement is needed to ensure that the screen information is properly displayed. However, we can do very thorough testing of the most critical aspects of the withdrawal module.

Requirements for the module are as follows:

1. Some accounts have a minimum balance requirement, indicated by Boolean variable `minflag`.
2. The bank allows all customers a basic overdraft protection amount, but for a fee, customers may purchase overdraft protection that exceeds the default.
3. If the account has a minimum balance, the withdrawal cannot reduce account balance below `(minimum balance - overdraft default)` unless overdraft protection is set for this account and the allowed overdraft amount for this account exceeds the default, in which case the balance cannot be reduced below `(minimum balance - overdraft protection amount)`.
4. No withdrawals may exceed the default limit (to keep the ATM from running out of cash), although some customers may have a withdrawal limit below this amount, such as minors who have an account with limits placed by parents.
5. The overdraft privilege can be used only once until the balance is made positive again.
6. Cards flagged as stolen are to be captured and logged in the hot card file. No withdrawal is allowed for a card flagged as stolen.

The module has these inputs from the user after the user is authorized by another module:

```
string num: the user card number
int amt: withdrawal amount requested
```

and these inputs from the system:

```
  int balance: user account balance
  boolean minflag: account has minimum balance
    requirement
  int min: account minimum balance
  boolean odflag: account has overdraft protection
  int odamt: overdraft protection amount,
  int oddefault: overdraft default
  boolean hot: card flagged as stolen
```

```
boolean limflag: withdrawal limit less than default
int limit: withdrawal limit for this account
int limdefault: withdrawal limit default
```

How should these requirements be translated into assertions and used in testing? Consider requirement 1: if `minflag` is set, then the balance before and after the withdrawal must be no less than the minimum balance amount. This could be translated directly into logic for assertions: `minflag => (balance >= min)`. If the assertion facility does not include logical implication, then the equivalent expression can be used, for example, in C syntax: `!minflag || balance >= min`.

However, we must also consider overdraft protection and withdrawal limits, so the assertion above is not adequate. Collecting conditions, we can develop assertions for each of the eight possible settings of `minflag`, `odflag`, and `limflag`. If there is a minimum balance requirement, no overdraft protection, and a withdrawal limit below the default, what is the relationship between balance and the other parameters?

```
minflag && !odflag && limflag
     => balance >= min - oddefault && amt <= limit
```

This relation must hold after the withdrawal, so to develop an assertion that must hold immediately before the withdrawal, substitute (balance – amt) for balance in the expression above:

```
balance - amt >= min - oddefault && amt <= limit
```

Assertions such as this would be placed *immediately* before the balance is modified, not at the beginning of the code for the withdrawal function. Code prior to the subtraction from balance should have ensured that properties encoded by assertions hold immediately before the subtraction; thus, any violation of the assertions indicates an error in the code (or possibly in the assertions!) that must be investigated. This is illustrated in Figure 11.1, where "wdl_init.c" and "wdl_final.c" are files containing assertions such as developed above.

Including the card number, there are 11 parameters for this module. We need to partition the inputs to determine what values to use in generating a covering array. Partitions should cover valid and invalid values, minimum and maximum for ranges, and values at and on either side of

```
while (!valid(acct)) {/* get account number input */}
if (amt  > lim ) { return ERROR;  }
else {
  if (odflag ) {
        if (amt  > balance  + odamt )
        { return ERROR;  }
  }
else {
  if (amt  > balance  + oddefault )
  {return ERROR;  }
  else {
        if (amt  > lim )
        { return ERROR;  }
  }
#include "wdl_init.c"
balance  -= amt ;
#include "wdl_final.c"
}
}
}
```

FIGURE 11.1 Withdrawal function code to be tested.

boundaries. The bank uses a check digit scheme for card numbers to detect errors such as digit transposition when numbers are entered manually. A simple partition could be as follows:

```
string acct: {valid, invalid}
int amt: {0, divisible by 20, not divisible by 20,
  max}
int balance: {0, negative, positive, max int}
int minflag: {T, F}
int min: {0, negative, positive, max int}
boolean odflag: {T, F}
int odamt: {0, negative, positive, max int}
int oddefault: {0, negative, positive, max int}
boolean hot: {T, F}
int acctlim: {0, negative, positive, max int}
int lim: {0, negative, positive, max int}
```

Using the equivalence classes above, this is thus a $2^4 4^7$ system, or 262,144 possible inputs. If values on either side of boundaries are used, the number of possible input combinations will be much larger, but using combinatorial methods, we can cover three-way or four-way combinations with only a few hundred tests.

11.2 STRONGER ASSERTION-BASED TESTING

The quality of assertion-based testing with combinatorial methods depends on the strength of assertions, in addition to *t*-way interaction strength.

While the method described in the previous section can be very effective in testing, notice that it will be inadequate for many problems, because basic assertion functions such as those provided in the C language library do not support important logic operators such as ∀ (*for all*) and ∃ (*for some*). Thus, expressing simple properties such as *S is sorted in ascending order* $= \forall i{:}0 \leq i < n - 1{:}S[i] \leq S[i+1]$ cannot be done without a good deal of additional coding. While it would be possible to add code to handle these problems in assertions, a better solution is to use an assertion language that is designed for the purpose and contains all the necessary features.

Tools such as Anna [121] for Ada, the JML [115] and iContract [95] for Java, and APP [166] or Nana [123] for C, can be used to introduce complex assertions, effectively embedding a specification within the code. An example of JML [218] can be seen in Figure 11.2. The assertions are annotated with "//@," to indicate statements that are input to the preprocessor. JML provides a collection of keywords, making it possible to specify the behavior of software and have the specifications checked as the program runs. Other assertion languages may use different keywords, but usually provide similar functionality. The basic run-time assertion checking features illustrated in the example are

- `//@ requires`: Defines a precondition, that is, a condition that must hold on entry to a module or section of code
- `//@ ensures`: Defines a postcondition, that is, a condition that must hold on exit from a module or section of code
- `//@ public invariant`: Defines an invariant, that is, a condition that must always hold
- `\old`: The value of the variable or expression on entry to the method
- `\result`: The return value of the method

JML and other assertion languages also provide features to make them easy to use for a specific programming language, and additional logic

```
public class BankingExample {

   public static final int MAX_BALANCE = 1000;
   private /*@ spec_public @*/ int balance;
   private /*@ spec_public @*/ boolean isLocked = false;

   //@ public invariant balance >= 0 && balance <= MAX_BALANCE;

   //@ assignable balance;
   //@ ensures balance == 0;
   public BankingExample() { balance = 0; }

   //@ requires 0 < amount && amount + balance < MAX_BALANCE;
   //@ assignable balance;
   //@ ensures balance == \old(balance + amount);
   public void credit(int amount) { balance += amount; }

   //@ requires 0 < amount && amount <= balance;
   //@ assignable balance;
   //@ ensures balance == \old(balance) - amount;
   public void debit(int amount) { balance -= amount; }

   //@ ensures isLocked == true;
   public void lockAccount() { isLocked = true; }

   //@   requires !isLocked;
   //@   ensures \result == balance;
   //@ also
   //@   requires isLocked;
   //@   signals_only BankingException;
   public /*@ pure @*/ int getBalance() throws BankingException {
       if (!isLocked) { return balance; }
       else { throw new BankingException(); }
   }
}
```

FIGURE 11.2 Toy bank module example in JML.

statements, such as the quantifiers `forall`, `exists` (*for some*), and logical implications: `a ==> b`, `a <== b`, `a <=> b`.

11.3 COST AND PRACTICAL CONSIDERATIONS

Assertions may be a cost-effective approach to test automation because they can be a simple extension of coding. In general, use of assertions is correlated with reduced error rates [113], but a very wide range of effectiveness results from variations in usage. In many applications, assertions are used in a very basic way, such as ensuring that null pointers are not passed to a function that will use them, or that parameters that may be used as divisors are nonzero.

More complex assertions can provide stronger assurance, but there are limits to their effectiveness. For example, invariants (properties that are expected to hold throughout a computation) generally cannot be assured without placing an assertion for every line of code. Since assertions must be executed to show the presence or absence of a property at some point, errors that prevent the assertion from being reached may not be detected. As an example, consider the code in Figure 11.1. If a coding error in the first few lines of the function prevents execution of the code at lines 15 and 17, the assertions will not be executed and it may be assumed that the test was passed. In this case, an ERROR return for the particular test case might trigger an investigation that would identify the faulty code, but this may not happen with other applications. Specialized assertion-checking languages such as JML can alleviate many of these problems by providing preprocessor statements to generate code that implements such complex checking without making the program difficult to read.

11.4 CHAPTER SUMMARY

Assertions are one of the easiest to use and most effective approaches to dealing with the oracle problem. Properties ranging from simple parameter checks to effectively embedded proofs can be encoded in assertions, but special language support is needed for the stronger forms of assurance. This support may be provided as language preprocessors, as in the case of Anna [121] and others. Placement within code is particularly important to assertion effectiveness [203,204], but if sufficiently strong assertions are embedded, the code becomes self-checking for important properties. With self-checking code, thousands of tests can be run at low cost in most cases, greatly improving the chances that faults will be detected.

REVIEW

Q: What is the difference between assertions and input validation code, as usually implemented?

A: Input validation code checks input supplied by the user or another function. Assertions encode properties that are expected to hold regardless of the input; they are intended to catch programming errors.

Q: Many popular languages such as C and Java provide an assertion function, but there are additional packages that can enhance the assertion capability for these languages. Why are these additional packages needed and what are some features that they provide?

A: Conventional assertion functions usually lack features such as logical quantifiers. Examples include *for all* and *exists* (*for some*) quantifiers, which are needed to express many properties needed in program specifications.

Q: Can assertions catch all errors in coding?

A: Not in most cases. Many complex functions are difficult to capture in assertions, particularly the limited assertion capabilities of most programming languages. However, assertions can often check for important core properties, such as numerical relationships between input and output values.

CHAPTER 12

Model-Based Testing

At first glance, the oracle problem for systems with extensive calculations and complex conditions may seem almost insoluble: how can we check its correctness without implementing equally complex software, whose correctness must also be checked, leading to an infinite string of verification exercises? Assertions and self-checking software can help (Chapter 11), but they are not always sufficient. In other cases, previous versions of the software may be available to check at least the old functionality, or the code may be implementing a formal standard (e.g., for network protocols or cryptography) and other implementations may exist to compare against. In most cases though, the software is doing something new, and we need to verify that it is working correctly for a large set of possible inputs. The difficulty of devising a set of complete tests with inputs and expected results is one of the reasons why *ad hoc* approaches such as "use cases" are widespread. Testers use formal or informal requirements to determine anticipated system uses, plus inputs and outputs for each such use, a slow and expensive way to develop a test oracle. To make thorough testing practical, more automated approaches are needed.

Model-based testing can provide strong assurance with a tradeoff of additional up-front time.

One of the most effective ways to produce test oracles is to use a model of the system under test and generate complete tests, including both input data and expected results directly from the model. We use the term *model* in the same way it would be used in other branches of engineering: the model incorporates aspects of the system that we want to study, but not every detail just as an aircraft model might be used in a wind tunnel to evaluate airflow but not all characteristics of a design. Models in software testing may be used to check calculations or performance, for example, but not other properties such as the location of a particular numeric value on a screen. (If it did include all details, the model would be equivalent to the system itself.) This chapter provides a step-by-step introduction to model-based automated generation of tests that provide combinatorial coverage. Procedures introduced in this tutorial will produce a set of complete tests, that is, input values with the expected output for each set of inputs.

In addition to the ACTS covering array generator (see Appendix C), we use NuSMV [43], a variant of the original symbolic model verifier (SMV) model checker [45]. NuSMV is freely available and was developed by Carnegie Mellon University, Instituto per la Ricerca Scientifica e Tecnolgica (IRST), University of Genova, and University of Trento. NuSMV can be installed on either UNIX/Linux or Windows systems running Cygwin. Links and instructions for downloading NuSMV are found at http://nusmv.fbk.eu/. The methods described in this chapter could of course be used with other model checkers as well, with some adaptation as needed for differences in capabilities of the different tools.

Also needed is a formal or semiformal specification of the system or subsystem under test (SUT). This can be in the form of a formal logic specification, but state transition tables, decision tables, pseudocode, or structured natural language can also be used, as long as the rules are unambiguous. The specification will be converted to SMV code, which provides a precise, machine-processable set of rules that can be used to generate tests.

12.1 OVERVIEW

To apply combinatorial testing, two tasks must be accomplished:

1. Using ACTS, construct a set of tests that will cover all *t*-way combinations of parameter values. The covering array specifies test data, where each row of the array can be regarded as a set of parameter values for an individual test (see Chapter 4).

2. Determine what output should be produced by the SUT for each set of input parameter values. The test data output from ACTS will be incorporated into SMV specifications that can be processed by the NuSMV model checker for this step. In many cases, the conversion to

SMV will be straightforward. The example in Section 12.2 illustrates a simple conversion of rules in the form "if *condition* then *action*" into the syntax used by the model checker. The model checker will instantiate the specification with parameter values from the covering array once for each test in the covering array. Because the model checker works to *disprove* claims, the resulting specification is evaluated against a claim that negates each specified result R_j to produce the expected result as a counterexample. Thus, the model checker evaluates claims in the following form: C_i => ~R_j, where C_i is a set of parameter values in row i of the covering array in the form $p_1 = v_{i1}$ & $p_2 = v_{i2}$ & · · · & $p_n = v_{in}$, and R_j is one of the possible results. The output of this step is a set of counterexamples that show how the SUT can reach the claimed result R_j from a given set of inputs.

The example in the following sections illustrates how these counterexamples are converted into tests. Other approaches to determining the correct output for each test can also be used. For example, in some cases, we can run a model checker in simulation mode, producing expected results directly rather than through a counterexample.

The completed tests can be used to validate correct operation of the system for interaction strengths up to some predetermined level t. Depending on the system type and level of effort, we may want to use pairwise ($t = 2$) or higher strength, up to $t = 6$-way interactions. We do not claim that this guarantees correctness of the system, as there may be failures triggered only by interaction strengths greater than t. In addition, some of the parameters are likely to have a large number of possible values, requiring that they be abstracted into equivalence classes. If the abstraction does not faithfully represent the range of values for a parameter, some flaws may not be detected by the equivalence class members used.

12.2 ACCESS CONTROL SYSTEM EXAMPLE

Here, we present a small example of a very simple access control system. The rules of the system are a simplified multilevel security system, given below, followed by a step-by-step construction of tests using a fully automated process.

Each subject (user) has a clearance level u_l, and each file has a classification level, f_l. Levels are given as 0, 1, or 2, which could represent levels such as confidential, secret, and top secret. A user u can read a file f if $u\_l \geq f\_l$ (the "no read up" rule), or write to a file if $f\_l \geq u\_l$ (the "no write down" rule).

Thus, a pseudocode representation of the access control rules is

```
if u_l >= f_l & act = rd then GRANT;
else if f_l >= u_l & act = wr then GRANT;
else DENY;
```

Tests produced will check that these rules are correctly implemented in a system.

12.3 SMV MODEL

This system is easily modeled in SMV as a simple two-state finite-state machine (FSM). The START state merely initializes the system (line 8, Figure 12.1), with the rule above used to evaluate access as either GRANT or DENY (lines 9–13). For example, line 9 represents the first line of the pseudocode above: in the current state (always START for this simple

```
MODULE main
1.       VAR
--Input parameters
2.       u_l:    0..2;           -- user level
3.       f_l:    0..2;           -- file level
4.       act:   {rd,wr};         -- action

--output parameter
5.       access: {START_, GRANT,DENY};

6.       ASSIGN
7.       init(access) := START_;
--if access is allowed under rules, then next state is GRANT
--else next state is DENY
8.       next(access) := case
9.       u_l >= f_l & act = rd : GRANT;
10.      f_l >= u_l & act = wr : GRANT;
11.      1 : DENY;
12.      esac;
13.      next(u_l) := u_l;
14.      next(f_l) := f_l;
15.      next(act) := act;

-- if user level is at or above file level then read is OK
SPEC AG ((u_l >= f_l & act = rd ) -> AX (access = GRANT));

-- if user level is at or below file level, then write is OK
SPEC AG ((f_l >= u_l & act = wr ) -> AX (access = GRANT));

-- if neither condition above is true, then DENY any action
SPEC AG (!( (u_l >= f_l & act = rd ) | (f_l >= u_l & act = wr ))
          -> AX (access = DENY));
```

FIGURE 12.1 SMV model of access control rules.

model), if $u\_l \geq f\_l$, then the next state is GRANT. Each line of the case statement is examined sequentially, as in a conventional programming language. Line 12 implements the "else DENY" rule, since the predicate "1" is always true. SPEC clauses given at the end of the model are simple "reflections" that duplicate the access control rules as temporal logic statements. They are thus trivially provable (see Figure 12.2), but we are interested in using them to generate tests rather than to prove properties of the system.

Separate documentation on SMV should be consulted to fully understand the syntax used, but specifications of the form "AG ((*predicate 1*) -> AX (*predicate 2*))" indicate essentially that for all paths (the "A" in "AG") for all states globally (the "G"), if *predicate 1* holds, then ("->") for all paths, in the next state (the "X" in "AX"), *predicate 2* will hold. In the next section, we will see how this specification can be used to produce complete tests, with test data input and the expected output for each set of input data.

If a property cannot be proved, the model checker produces a counterexample, giving inputs and paths that lead to the violation.

Model checkers can be used to perform a variety of valuable functions because they make it possible to evaluate whether certain properties are true of the system model. Conceptually, the model checker can be viewed as exploring all states of a system model to determine if a property claimed in a SPEC statement is true. If the statement can be proved true for the given model, the model checker reports this fact. What makes a model checker particularly valuable for many applications, though, is that if the statement is false, the model checker not only reports this but also provides a "counterexample" showing how the claim in the SPEC statement can be shown false. The counterexample will include input data values and a trace of system states that lead to a result contrary to the SPEC claim (Figure 12.2). In the process described in this section, the input data values will be the covering array generated by ACTS.

```
-- specification AG((u_l >= f_l & act = rd) -> AX access = GRANT)
    is true
-- specification AG((f_l >= u_l & act = wr) -> AX access = GRANT)
    is true
-- specification AG(!((u_l >= f_l & act = rd)|(f_l >= u_l & act = wr))
                                        -> AX access = DENY)  is true
```

FIGURE 12.2 NuSMV output.

For advanced uses in test generation, this counterexample generation capability is very useful for proving properties such as liveness (absence of deadlock) that are difficult to ensure through testing. In this tutorial, however, we will simply use the model checker to determine whether a particular input data set makes a SPEC claim true or false. That is, we will enter claims that particular results can be reached for a given set of input data values, and the model checker will tell us if the claim is true or false. This gives us the ability to match every set of input test data with the result that the system should produce for that set of input data.

The model checker thus automates the work that normally must be done by a human tester—determining what the correct output should be for each set of input data. In some cases, we may have a "reference implementation," that is, an implementation of the functions that we are testing that is assumed to be correct. This happens, for example, in conformance testing for protocols, where many vendors implement their own software for the protocol and submit it to a test lab for comparison with an existing implementation of the protocol. In this case, the reference implementation could be used for determining the expected output, instead of the model checker. Of course, before this can happen, the reference implementation itself must be thoroughly tested before it can be the gold standard for testing other products. The method we describe here may be needed to produce tests for the original reference implementation.

Checking the properties in the SPEC statements shows that they match the access control rules as implemented in the FSM, as expected (see Figure 12.2). In other words, the claims we made about the state machine in the SPEC clauses can be proven. This step is used to check that the SPEC claims are valid for the model defined previously. If NuSMV is unable to prove one of the SPECs, then either the spec or the model is incorrect. This problem must be resolved before continuing with the test generation process. Once the model is correct and SPEC claims have been shown to be valid for the model, counterexamples can be produced that will be turned into test cases, by which we mean a set of test inputs with the expected result for these inputs. In other words, ACTS is used to generate tests, then the model checker determines expected results for each test.

12.4 INTEGRATING COMBINATORIAL TESTS INTO THE MODEL

We will compute covering arrays that give all t-way combinations, with interaction strength = 2 for this example. This section describes the use of ACTS as a standalone command line tool, using a text file input (see Appendix D). The first step is to define the parameters and their values in a system definition file that will be used as input to ACTS. Call this file "in.txt," with the following format:

```
[System]
[Parameter]
  u_l: 0,1,2
  f_l: 0,1,2
  act: rd,wr
[Relation]
[Constraint]
[Misc]
```

For this application, the [Parameter] section of the file is all that is needed. Other tags refer to advanced functions that will be explained in other documents. After the system definition file is saved, run ACTS as shown below:

```
java -Ddoi=2 -jar acts_cmd.jar ActsConsoleManager
in.txt out.txt
```

The "-Ddoi = 2" argument sets the interaction strength (degree of interaction) for the covering array that we want ACTS to compute. In this case, we are using simple two-way, or pairwise, interactions. (For a system with more parameters, we would use a higher-strength interaction, but with only three parameters, three-way interaction would be equivalent to exhaustive testing.) ACTS produces the output shown in Figure 12.3.

Each test configuration defines a set of values for the input parameters u_l, f_l, and act. The complete test set ensures that all two-way combinations of parameter values have been covered. If we had a larger number of parameters, we could produce test configurations that cover all three-way, four-way, and other combinations. ACTS may output "don't care" for some parameter values. This means that any legitimate value for that parameter can be used and the full set of configurations will still cover all *t*-way combinations. Since "don't care" is not normally an acceptable input for programs being tested, a random value for that parameter is substituted before using the covering array to produce tests.

The next step is to assign values from the covering array to parameters used in the model. For each test, we claim that the expected result

```
Number of parameters: 3
  Maximum number of values per parameter: 3
  Number of configurations: 9
  -------------------------------------
  Configuration #1:
  1 = u_l=0
  2 = f_l=0
  3 = act=rd
  -------------------------------------
  Configuration #2:
  1 = u_l=0
  2 = f_l=1
  3 = act=wr
  -------------------------------------
  Configuration #3:
  1 = u_l=0
  2 = f_l=2
  3 = act=rd
  -------------------------------------
  Configuration #4:
  1 = u_l=1
  2 = f_l=0
  3 = act=wr
  -------------------------------------
  Configuration #5:
  1 = u_l=1
  2 = f_l=1
  3 = act=rd
  -------------------------------------
  Configuration #6:
  1 = u_l=1
  2 = f_l=2
  3 = act=wr
  -------------------------------------
  Configuration #7:
  1 = u_l=2
  2 = f_l=0
  3 = act=rd
  -------------------------------------
  Configuration #8:
  1 = u_l=2
  2 = f_l=1
  3 = act=wr
  -------------------------------------
  Configuration #9:
  1 = u_l=2
  2 = f_l=2
  3 = (don't care)
```

FIGURE 12.3 ACTS output.

will not occur. The model checker determines combinations that would disprove these claims, outputting these as counterexamples. Each counterexample can then be converted to a test with known expected result. Every test from the ACTS tool is used, with the model checker supplying expected results for each test. (Note that the trivially provable positive claims have been commented out. Here, we are concerned with producing counterexamples.)

Recall the structure introduced in Section 12.1: C_i => ~R_j. Here, C_i is the set of parameter values from the covering array. For example, for configuration #1 in Figure 12.3:

```
u_l=0 & f_l=0 & act=rd
```

As can be seen below, for each of the nine configurations in the covering array, we create a SPEC claim of the form:

```
SPEC AG((<covering array values>) -> AX
  !(access=<result>));
```

This process is repeated for each possible result, in this case, either "GRANT" or "DENY," so we have nine claims for each of the two results. The model checker is able to determine, using the model defined in Section 12.2, which result is the correct one for each set of input values, producing a total of nine tests.

Excerpt:

```
...
– reflection of the assign for access
– SPEC AG ((u_l>= f_l & act=rd) -> AX
  (access=GRANT));
– SPEC AG ((f_l>= u_l & act=wr) -> AX
  (access=GRANT));
– SPEC AG (!((u_l>= f_l & act=rd) | (f_l>= u_l &
  act=wr)) -> AX (access=DENY));
SPEC AG((u_l=0 & f_l=0 & act=rd) -> AX
  !(access=GRANT));
SPEC AG((u_l=0 & f_l=1 & act=wr) -> AX
  !(access=GRANT));
SPEC AG((u_l=0 & f_l=2 & act=rd) -> AX
  !(access=GRANT));
SPEC AG((u_l=1 & f_l=0 & act=wr) -> AX
  !(access=GRANT));
```

```
SPEC AG((u_l=1 & f_l=1 & act=rd) -> AX
  !(access=GRANT));
SPEC AG((u_l=1 & f_l=2 & act=wr) -> AX
  !(access=GRANT));
SPEC AG((u_l=2 & f_l=0 & act=rd) -> AX
  !(access=GRANT));
SPEC AG((u_l=2 & f_l=1 & act=wr) -> AX
  !(access=GRANT));
SPEC AG((u_l=2 & f_l=2 & act=rd) -> AX
  !(access=GRANT));
SPEC AG((u_l=0 & f_l=0 & act=rd) -> AX
  !(access=DENY));
SPEC AG((u_l=0 & f_l=1 & act=wr) -> AX
  !(access=DENY));
SPEC AG((u_l=0 & f_l=2 & act=rd) -> AX
  !(access=DENY));
SPEC AG((u_l=1 & f_l=0 & act=wr) -> AX
  !(access=DENY));
SPEC AG((u_l=1 & f_l=1 & act=rd) -> AX
  !(access=DENY));
SPEC AG((u_l=1 & f_l=2 & act=wr) -> AX
  !(access=DENY));
SPEC AG((u_l=2 & f_l=0 & act=rd) -> AX
  !(access=DENY));
SPEC AG((u_l=2 & f_l=1 & act=wr) -> AX
  !(access=DENY));
SPEC AG((u_l=2 & f_l=2 & act=rd) -> AX
  !(access=DENY));
```

12.5 GENERATING TESTS FROM COUNTEREXAMPLES

Counterexamples from the model checker can be postprocessed into complete tests, with inputs and expected output for each.

NuSMV produces counterexamples where the input values would disprove the claims specified in the previous section. Each of these counterexamples is thus a set of test data that would have the expected result of GRANT or DENY.

For each SPEC claim, if this set of values cannot in fact lead to the particular result R_j, the model checker indicates that this is true. For example, for the configuration below, the claim that access will not be granted is true, because

the user's clearance level (u _ l = 0) is below the file's level (f _ l = 2):
```
– specification AG (((u_l = 0 & f_l = 2) & act = rd) -> AX
!(access = GRANT)) is true
```

If the claim is false, the model checker indicates this and provides a trace of parameter input values and states that will prove it is false. In effect, this is a complete test case, that is, a set of parameter values and expected result. It is then simple to map these values into complete test cases in the syntax needed for the system under test.

Excerpt from NuSMV output:

```
– specification AG (((u_l = 0 & f_l = 0) & act = rd) ->
  AX
access = GRANT)) is false
– as demonstrated by the following execution
  sequence
Trace Description: CTL Counterexample
Trace Type: Counterexample
-> State: 1.1 <-
  u_l = 0
  f_l = 0
  act = rd
access = START_
-> Input: 1.2 <-
-> State: 1.2 <-
access = GRANT
```

The model checker finds that six of the input parameter configurations produce a result of GRANT and three produce a DENY result, so at the completion of this step, we have successfully matched up each input parameter configuration with the result that should be produced by the SUT.

We now strip out the parameter names and values, giving tests that can be applied to the system under test. This can be accomplished using a variety of methods; a simple script was used. The test inputs and expected results produced are shown below:

```
u_l = 0 & f_l = 0 & act = rd -> access = GRANT
u_l = 0 & f_l = 1 & act = wr -> access = GRANT
u_l = 1 & f_l = 1 & act = rd -> access = GRANT
u_l = 1 & f_l = 2 & act = wr -> access = GRANT
u_l = 2 & f_l = 0 & act = rd -> access = GRANT
```

```
u_l=2 & f_l=2 & act=rd -> access=GRANT
u_l=0 & f_l=2 & act=rd -> access=DENY
u_l=1 & f_l=0 & act=wr -> access=DENY
u_l=2 & f_l=1 & act=wr -> access=DENY
```

These test definitions can now be postprocessed using simple scripts written in PERL, Python, or similar tool to produce a test harness that will execute the SUT with each input and check the results. While tests for this trivial example could easily have been constructed manually, the procedures introduced in this tutorial can be, and have been, used to produce tens of thousands of complete test cases in a few minutes, once the SMV model has been defined for the SUT [102,103].

12.6 COST AND PRACTICAL CONSIDERATIONS

Model-based testing can reduce overall cost because of the tradeoffs involved.

Model-based test generation trades up-front analysis and specification time against the cost of greater human interaction for analyzing test results. The model or formal specification may be costly to produce, but once it is available, large numbers of tests can be generated, executed, and analyzed without human intervention. This can be an enormous cost saving, since testing usually requires 50% or more of the software development budget. For example, suppose a $100,000 development project expects to spend $50,000 on testing, because of the staff time required to code and run tests, and analyze results. If a formal model can be created for $20,000, complete tests generated and analyzed automatically, with another $10,000 for a smaller number of human-involved tests and analysis, then the project will save 20%. One tradeoff for this savings is the requirement for staff with skills in formal methods, but in some cases, this approach may be practical and highly cost-effective.

One nice property of the model checking approach described in this chapter is that test case generation can be run in parallel. For each test row of the covering array, we run the model checker to determine the expected results for the inputs given by that row, and model checker runs are independent of each other. Thus, this task falls into the class of parallelization problems known as "embarrassingly parallel"; for *N* covering array rows, we can assign up to *N* processors. With the widespread availability of cloud and cluster systems, test generation can run very quickly. In most cases, test execution can be run in parallel also, although we may be limited by practical concerns such as availability of specialized hardware.

12.7 CHAPTER SUMMARY

1. The oracle problem must be solved for any test methodology, and it is particularly important for thorough testing that produces a large number of test cases. One approach to determining expected results for each test input is to use a model of the system that can be simulated or analyzed to compute output for each input.

2. Model checkers can be used to solve the oracle problem because whenever a specified property for a model does not hold, the model checker generates a counterexample. The counterexample can be postprocessed into a complete working test harness that executes all tests from the covering array and checks results.

3. Several approaches are possible for integrating combinatorial testing with model checkers, but some present practical problems. The method reported in this chapter can be used to generate full combinatorial test suites, with expected results for each test, in a cost-effective way.

REVIEW

Q: What is meant by a *test oracle*?

A: The test oracle is a function that determines if a particular result is correct for a given input.

Q: What are some limitations of using model-based testing?

A: A model does not duplicate all functions of a system, only the ones that the model builder chooses to include. Typically, these will be major results that are critical to system operation, but may exclude less essential functionality. For example, a model for an e-commerce application may be designed to test financial computations, but leave out issues such as text formatting.

Q: What does the term *counterexample* mean in the context of model checking?

A: A counterexample is a demonstration that some claim made about the specification does not hold in all cases. The counterexample provides a set of inputs and steps in the modeled system for which the claimed property does not hold. For example, in a continuously running real-time system, we may make a claim that no deadlocks are possible. If a counterexample can be found by the model checker, it indicates that there is a problem because the system may become deadlocked in some cases.

Q: Why is the counterexample-finding ability of model checkers important for creating a test oracle?

A: The model checker's counterexamples can specify expected results for given inputs according to the model of the system under test.

Q: Model checking is one way of generating expected output for a given set of inputs. What are some other ways?

A: Manual computation or simulation can also be used. If the SUT is a new implementation (e.g., on different hardware) of an existing function, the previous code may be used as a test oracle. This occurs with implementations of standards such as protocols, text formatting as in web browsers, and cryptographic libraries, among others.

CHAPTER 13

Fault Localization

In combinatorial testing, detection of faults in the SUT (system under test) requires two steps: the first step from the pass/fail data on the tests conducted identifies the particular combination or combinations that led to the failure; after the failure-inducing combinations are identified, search for the actual faults (mistakes) in the SUT that are associated with the failure-inducing combinations. The second step requires access to the source code of the SUT. In this chapter, we address only the first step of identifying failure-inducing combinations. The problem of fault localization, identifying such combination(s), is an area of active research. The discussion in this chapter assumes that systems are deterministic, such that a particular input always generates the same output.

At first glance, fault localization may not appear to be a difficult problem, and in many cases, it will not be difficult, but we want to automate the process as much as possible. To understand the size of the problem, consider a module that has 20 input parameters. A set of 3-way covering tests passes 100% but several tests derived from a 4-way covering array result in failure. (Therefore, at least four parameter values are involved in triggering the failure. It is also possible that a 5-way or higher combination caused the failure since any set of t-way tests also includes $(t + 1)$-way and higher-strength combinations as well. Indeed, a test suite of $t + k$ parameters (where $k \geq 0$) consists of a number of $(t + k)$-way combinations.) A test with 20 input parameters has $C(20, 4) = 4845$ 4-way combinations; yet, presumably only one (or just a few) of these combinations triggered the failure. How can the failure-triggering combinations be identified with thousands of possible combinations to consider?

Several methods have been developed to locate faults directly from test results, which are not restricted to combinatorial tests. Among the better known methods are delta debugging [222]. For combinatorial testing, the methods have been categorized as adaptive, in which test sets are revised and rerun based on the previous tests, or nonadaptive [213]. One nonadaptive method [56] detects faults if the number of faults is no more than a prespecified value d, and all faults are triggered by t-way combinations of values. A number of adaptive methods have been proposed, for example, Refs. [74,144,174,178,213]. Generally, adaptive methods require fewer assumptions, and thus, they may be more broadly applicable than nonadaptive approaches. This chapter introduces the fault localization problem, describes the basic framework for fault localization using adaptive testing, and discusses how some of these approaches can be implemented.

13.1 FAULT LOCALIZATION PROCESS

The analysis presented here applies to a deterministic system, in which a particular set of input values always results in the same processing and outputs. Let $P = \{t\text{-way combinations in passing tests}\}$, $F = \{t\text{-way combinations in failing tests}\}$, and $C = \{t\text{-way failure-triggering combinations}\}$. A failure-triggering combination is a combination that causes any test containing this combination to fail. Then, the set difference $F \backslash P$ (t-way combinations in failing tests that are not in any passing tests) must contain the fault-triggering combinations C because if any of those combinations in C were in P, then the test would have failed. So, $C \subseteq F \backslash P$, as shown in Figure 13.1. Note that C could be empty.

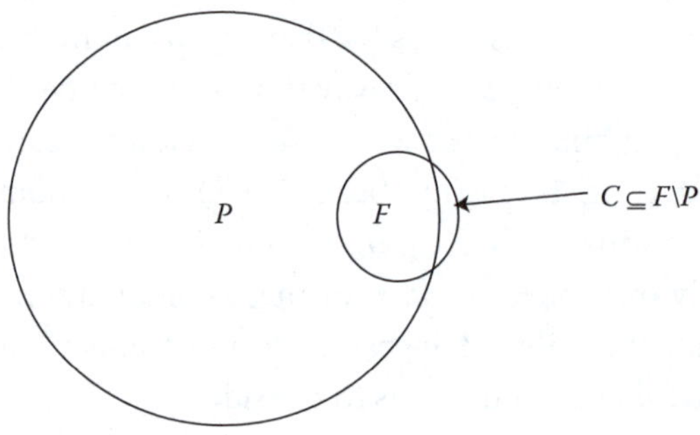

FIGURE 13.1 Combinations in failing tests but not in passing tests.

Continuing with the analysis in this manner, some properties become apparent. Since sets F and P are the result of partitioning the original covering array, all t-way combinations are present in either passing or failing tests, or both. Using the properties above, simple set operations can identify a small number of suspect t-way combinations that must contain the failure-triggering combination, if tests from a full t-way covering array are run on the SUT. Once tests have been run, the test set is augmented with additional tests that are run, and passing tests are added to P. The choice of additional tests can be made in a variety of ways, as discussed below. As combinations from new passing tests are added to P, the number of combinations in $F \backslash P$ continues to be reduced only until fault-triggering combinations are contained in $F \backslash P$.

13.1.1 Analyzing Combinations

Suppose that a particular combination c triggers a failure if whenever c is contained in some test, the test fails. Combinations such as c are referred to as failure-triggering combinations. A variety of heuristics [108] exists for identifying the failure-triggering combination(s) C.

- *Elimination*: For a deterministic system, $F \backslash P$ must contain the failure-triggering combinations C because if any of those combinations in C were in P, then the test would have failed.
- *Interaction level lower bound*: If all t-way coverage tests pass, then a t-way or lower strength combination did not cause the failure. The failure must have been caused by a k-way combination, for some $k > t$. Note that the converse is not necessarily true: if some t-way coverage test fails, we cannot conclude that a t-way test caused the failure because any t-way test set contains some k-way combinations, for $k > t$.
- *Value dependence*: If tests in $F(t)$ cover all values for a t-way parameter combination c, then the failure is probably independent of the values of c; that is, c is not a t-way failure-triggering combination. The exception for this rule is the rare case where all value combinations of c are failure triggering.
- *Ranking*: Combinations can be ranked using a variety of metrics, such as using a ratio of the number of failing tests containing a particular combination over the total number of failing tests [74]. The intuition here is that combinations that occur more often in failing tests than other combinations are more likely to be problematic.

13.1.2 New Test Generation

Several approaches have been proposed for generating new tests in the adaptive fault location process, but the basic idea is to find tests that are most effective in reducing the size of the set of suspicious combinations. Since combinations that appear in a passing test cannot be failure inducing, it is desirable to generate new tests that are likely to pass and are likely to contain some combinations that are currently suspicious.

13.1.2.1 Alternate Value

Let T_f be a failing test. For each of the $i = 1,\ldots,n$ parameters, create one new test for each parameter i in which the value of parameter i is changed to a different value and all other parameter values are held constant. The new test is likely to contain some combinations that are also contained in T_f. If the new test passes, then these combinations can be removed from the set of suspicious combinations. For example, if we have five binary parameters, $T_f = 01011$, create 11011, 00011, 01111, 01001, and 01010. This procedure generates kn new tests, where k = number of failing tests, n = number of parameters.

13.1.2.2 Base Choice

This method extends the alternate value scheme to produce tests with each alternate value instead of a single one. For each of the $i = 1,\ldots,n$ parameters, create one new test for each other possible value v of each parameter i in which the value of parameter i is changed to v and all other parameter values are held constant. For example, if each parameter has three possible values, 0, 1, and 2, and $T_f = 10212$, create 20212, 00212, 11212, 12212, and so on. The base choice procedure creates $kn(v - 1)$ new tests, where k = number of tests identified in Step 2, n = number of parameters, v = number of values per parameter. If parameters have different numbers of values, v_i values for parameter i, then $\Sigma_{i=1,\ldots,n}(v_i - 1)$ new tests are created.

This procedure reduces the set of suspicious combinations and can identify failure-triggering combinations where each test in F contains at most one fault-triggering combination. It works by moving nonfault-triggering combinations in a failing test T_f to the set P^+. Using the base choice procedure, we alter one parameter at a time in the failing test. If the fault-triggering combination c contains t parameters, then the base choice procedure is likely to produce at least one test T' from T_f for which one of the parameter values of c has changed such that T' passes. Identify the parameter values in the fault-triggering combination c as c_j, $j = 1,\ldots,t$. Any combination from T_f that was originally in F either contains no parameters from c, or contains one or more parameters of c_j. If it contains no parameters from c, then every new test created from T_f moves it to P. This property holds as long as the new test produced by changing a value in c does not introduce any new failure-triggering combinations. This is likely to be true if the software being tested contains few errors, but it is not guaranteed.

13.2 LOCATING FAULTS: EXAMPLE

In the preceding discussion, we assumed that a particular combination c triggers or causes a failure if whenever c is contained in some test x, test x fails. However, in many cases, the presence of a particular combination may trigger a failure, but is not guaranteed to do so (see discussion of interaction level lower bound above). Consider the following:

```
1.  p(int a, int b, int c, int d, int e) {
2.  If (a && b)           return 1;
3.  Else if (c && d)      return 2;
4.  Else if (e)           return 3;
5.  Else                  return 4;
6.  }
```

If line 3 is incorrectly implemented as “return 7” instead of “return 2,” then $p(1,1,1,1,0) = 1$ because “a && b” evaluates to 1, masking the error, but $p(0,1,1,1,0)$ will detect the error. A complete 3-way covering test set will detect the error because it must include at least one test with values (0,*,1,1,*) and one test with values (*,0,1,1,*). That is, a 3-way covering array must include combinations that make (a && b) false and make (c && d) true, allowing line 7 to be executed. Table 13.1 shows tests for this example for $t = 2$, 3, and 4. The failing tests are underlined.

A 2-way test *may* detect the error since “c && d” is the condition necessary, but this will only occur if line 3 is reached, which requires either $a = 0$ or $b = 0$. In the example of the test set, this occurs with the second test. So, in this case, a full 2-way test set has detected the error, and the heuristics above for 2-way combinations will find that tests with $c = 1$ and $d = 1$ occur in both P and F. In this case, debugging may identify $c = 1$, $d = 1$ as a combination that triggers the failure, but automated analysis

TABLE 13.1 Tests for Fault Localization Example

1-Way Tests	2-Way Tests	3-Way Tests	4-Way Tests
0,0,0,0,0	0,0,0,0,0	0,0,0,0,0	0,0,0,0,0
1,1,1,1,1	0,1,1,1,1	0,0,1,1,1	0,0,0,1,1
	1,0,1,0,1	0,1,0,1,0	0,0,1,0,1
	1,1,0,1,0	0,1,1,0,1	0,0,1,1,0
	1,1,1,0,0	1,0,0,1,1	0,1,0,0,1
	1,0,0,1,1	1,0,1,0,0	0,1,0,1,0
		0,1,1,0,1	0,1,1,1,0
		1,1,1,1,0	0,1,1,1,1
		0,0,1,1,0	1,0,0,0,1
		1,1,0,0,0	1,0,0,1,0
		0,0,0,0,1	1,0,1,0,0
		1,1,1,1,1	1,0,1,1,1
		0,1,1,1,0	1,1,0,0,0
			1,1,0,1,1
			1,1,1,0,1
			1,1,1,1,0

using the heuristics will find two 3-way combinations that occur in failing tests but not in passing tests: $a = 0$, $c = 1$, $d = 1$ and $b = 0$, $c = 1$, $d = 1$. As Table 13.1 illustrates, in most cases, we will find more than one combination identified as possible causes of failure.

The heuristics described in the preceding section can be applied to combinations in the failed tests to identify possible failure-triggering combinations, shown in Table 13.2.

- The 1-way tests do not detect any failures, but the 2-way tests do, so, $t = 2$ is a lower bound for the interaction level needed to detect a failure for this example program.
- The value dependence rule applies to combination "be"—since all four possible values for this combination occur in failing tests; the failure is probably independent of combination be. In other words, we do not consider the pair "be" to be a cause of failure because it does not matter what value this pair has. Every test must have some value for these parameters.
- The elimination rule can be applied to determine that there are no 1-way or 2-way combinations that do not appear in both passing and failing tests. The results for 3- and 4-way combinations are shown in Table 13.3. These results were produced by an analysis tool

TABLE 13.2 Combinations in Failing Tests

$t = 2$	**ab**	**ac**	**ad**	**ae**	**bc**	**bd**	**be**	**cd**	**ce**	**de**
	01	01	01	01	11	11	11	11	11	11
	00	11	11	00	01	01	01		10	10
	10			11			00			
							10			
$t = 3$	**abc**	**abd**	**abe**	**acd**	**ace**	**ade**	**bcd**	**bce**	**bde**	**cde**
	011	011	011	011	011	011	111	111	111	111
	001	001	001	111	010	010	011	011	011	110
	101	101	000		111	111		010	010	
			101					110	110	
			010							
$t = 4$	**abcd**	**abce**	**abde**	**bcde**						
	0111	0111	0111	1111						
	0011	0011	0011	0111						
	1011	0010	0010	0110						
		1011	1011	1110						
		0110	0110							

TABLE 13.3 3- and 4-Way Combinations in $F \backslash P$

1:3-way 0,2,3 = 0,1,1	1:4-way 0,1,2,3 = 0,0,1,1
2:3-way 0,2,3 = 0,1,1	2:4-way 0,1,2,3 = 0,0,1,1
3:3-way 0,2,3 = 0,1,1	3:4-way 0,1,2,3 = 0,1,1,1
4:3-way 0,2,3 = 0,1,1	4:4-way 0,1,2,3 = 0,1,1,1
1:3-way 1,2,3 = 0,1,1	5:4-way 0,1,2,3 = 1,0,1,1
2:3-way 1,2,3 = 0,1,1	1:4-way 0,1,2,4 = 0,0,1,0
5:3-way 1,2,3 = 0,1,1	1:4-way 0,1,3,4 = 0,0,1,0
	4:4-way 0,1,3,4 = 0,1,1,1
	1:4-way 0,2,3,4 = 0,1,1,0
	2:4-way 0,2,3,4 = 0,1,1,1
	3:4-way 0,2,3,4 = 0,1,1,0
	4:4-way 0,2,3,4 = 0,1,1,1
	1:4-way 1,2,3,4 = 0,1,1,0
	2:4-way 1,2,3,4 = 0,1,1,1
	5:4-way 1,2,3,4 = 0,1,1,1

that outputs in the format <test number> : <*t* level> <parameter numbers> = <parameter values>. Two different 3-way combinations are identified: $a = 0$, $c = 1$, $d = 1$ and $b = 0$, $c = 1$, $d = 1$. A large number of 4-way combinations are also identified, but we can use the interaction continuity rule to show that one of the two 3-way combinations occurs in all the 4-way failing tests. Therefore, we can conclude that covering all 3-way parameter interactions would detect the error.

The situation is more complex with continuous variables. If, for example, a failure-related branch is taken any time $x > 100$, $y = 3$, $z < 1000$, there may be many combinations implicated in the failure. Analysis will show that [$x = 200$, $y = 3$, $z = 120$], [$x = 201$, $y = 3$, $z = 119$], [$x = 999$, $y = 3$, $z = 999$], [$x = 101$, $y = 3$, $z = 0$], and [$x = 200$, $y = 3$, $z = 0$] are all combinations that trigger the failure. With more than three input parameters, there may be dozens or hundreds of failure-triggering combinations, even though there is most likely a single point in the code that is in error.

13.2.1 Generating New Tests

Assume that a system has binary variables a through e, and the 2-way combination $a = 0$, $b = 1$ triggers a failure.

A covering array for this system is

0,0,0,0,0
0,1,1,1,1
1,0,1,0,1
1,1,0,1,0
1,1,1,0,0
0,0,0,1,1

The row containing $a = 0$, $b = 1$ is 0,1,1,1,1, which becomes set F. Set P is combinations from

0,0,0,0,0
1,0,1,0,1
1,1,0,1,0
1,1,1,0,0
0,0,0,1,1

Now, generate the base choice tests, as shown below. Note that not all the base choice tests may be necessary since we only need to change the fault-triggering combination c such that T' passes; so, it may not be necessary to run all these tests before hitting a new test that passes, and then recomputing the set difference. An alternative is to generate the supplemental tests one at a time and to do the computation $F \backslash P$ after each test.

The choice of procedures depends on the trade-off between test execution time and effectiveness.

The additional tests and results are

1,1,1,1,1 (pass)
0,0,1,1,1 (pass)
0,1,0,1,1 (fail)
0,1,1,0,1 (fail)
0,1,1,1,0 (fail)

Using the new information from the base choice tests, we have two additional passing tests: 1,1,1,1,1 and 0,0,1,1,1. Adding these to the previous passing set P to produce P^+, we have

0,0,1,1,1
0,0,0,0,0
1,0,1,0,1
1,1,0,1,0
1,1,1,0,0
0,0,0,1,1
1,1,1,1,1

Then computing $F \backslash P^+ = c = (ab = 01)$, thus correctly identifying the fault. Note that we would need at most $n - 1$ new base choice tests to find a parameter involved in failure (at worst, the last two parameters would be the fault-triggering combination). Thus, this method requires a total of $\Sigma_{i=1,\ldots,n} (v_i - 1)$ new tests to augment the existing set.

13.3 COST AND PRACTICAL CONSIDERATIONS

As shown in the example above, it is a nontrivial matter to determine the failure-triggering combination(s) from test results alone. In black-box testing situations where there is no access to source code, the methods described in this section are useful in understanding the behavior of a program with respect to its input. When source code is available, the methods described in this section can help to reduce the scope of the code that needs to be inspected by conventional debugging techniques. The tools to implement these methods have been developed and some tools are available from the ACTS project site. The more advanced approaches under development include methods and tools to locate source code errors from t-way test results [74].

13.4 CHAPTER SUMMARY

A t-way combinatorial test set is designed to detect failures that are triggered by combinations involving no more than t parameters. Assume that we have executed a t-way test set and some tests have failed. A natural question to ask is: What combinations have caused these failures? Identifying such combinations can facilitate the debugging effort, for example, by reducing the scope of the code that needs to be inspected.

A failure-triggering combination is a combination of parameter values such that all test cases containing this combination fail. Failure-triggering combinations can be identified in two major steps. First, simple set analysis can be carried out on the testing result of a combinatorial test set to identify a set of suspect combinations. The key observation is that combinations that appear in a passing test cannot be failure triggering. Second, additional tests can be generated to refine the set of suspect combinations. Several approaches have been developed that generate new tests that are likely to pass and are likely to contain suspect combinations. Each time a new test passes, all the suspect combinations contained in the new test are removed from the set of suspect combinations.

REVIEW

Q: What are the two common approaches for fault localization?

A: Adaptive and nonadaptive. Nonadaptive schemes attempt to locate failure-triggering combinations from a particular set of test results, whereas adaptive approaches use an iterative process of identifying suspect combinations, and then they generate new tests to confirm or eliminate combinations from the set of possible failure-triggering combinations.

Q: What are some of the trade-offs between adaptive and nonadaptive methods?

A: Nonadaptive methods can clearly be faster for some problems, but they do not apply in all circumstances. Adaptive approaches require fewer assumptions but may require many additional tests to locate the problem.

CHAPTER 14

Evolution from Design of Experiments

Combinatorial testing of software can be viewed as an adaptation of design of experiment (DOE) methods for industrial applications. Although the general outline of DOE and combinatorial software testing is similar, the application of DOE to software is not a simple translation of DOE methods to a different domain. The unique characteristics of software have required significant advances, and understanding how these developments occurred can deepen our understanding of combinatorial testing.

14.1 BACKGROUND

As discussed in the previous chapters, combinatorial testing is a type of dynamic testing in which distinct (but related) test factors are specified. Dynamic testing means that the software system is exercised (run) to obtain information about it. The test factors may have all possible values on a continuous scale or on a discrete scale. In either case, a set of a few discrete possible values is selected by equivalence partitioning, by boundary value analysis, or by expert judgment. Thus, for each test factor, a small set of discrete test settings is specified. Each test case is then expressed as a combination of one test setting for every test factor (one value for each parameter) [4,125,128,194]. Suppose there are k test factors of which k_1 have v_1 test settings each, k_2 have v_2 test settings each, ..., and k_n have v_n test settings each. A test case consists of a combination (k-tuple) of k test settings one for each test factor, where $k_1 + k_2 + \cdots + k_n = k$. The number of different test cases possible is the product of all k test values. For example, if $k_1 = 3$, $v_1 = 3$, $k_2 = 4$, $v_2 = 4$, $k_3 = 2$, and $v_3 = 5$, then we have $3 + 4 + 2 = 9$ test factors (3 with three test settings each, 4 with four test settings each, and 2 with five test settings each) and the number of possible test cases is $3^3 4^4 5^2 = 172{,}800$. The exponential expression $3^3 4^4 5^2$ represents the combinatorial test structure of the k test factors and its expanded form 172,800 is the number of possible test cases. In many practical applications, the number of possible test cases is too large to test all of them. Combinatorial testing methods are used to determine a small set of test cases (test suite), which exercise the test settings of each test factor and their combinations with the objective that the test suite will cover the combinations for which the SUT could possibly fail.

The test factors and their test settings, and the expected behavior of the SUT are determined from the documents of requirements, descriptions of system implementations and internal operations, and all other available information about the SUT. The choice of test factors and their test settings defines and limits the scope of the combinations that are tested. Clearly, faults involving factors and settings that are not chosen for testing may not be exercised and revealed [194]. Methods for specification of test factors and their test settings are largely application domain specific and a subject of continuing research [4]. Chapters 3 through 6 include approaches to this topic in the context of combinatorial testing.

Orthogonal arrays (OAs) are mathematical objects that are used as templates for designs (plans) of industrial experiments. Their key property is that every t-way combination appears *exactly the same number of times.* Covering arrays are generalizations of OAs that are especially suited for software testing. They require each t-way combination to appear *at least once.* Covering arrays offer significant advantages over OAs for testing software. Combinatorial testing began as pairwise testing in which first OAs and then covering arrays were used to make sure that all test settings of each factor and all pairs of the test settings were tested (excluding invalid test cases). In Section 14.2, we discuss pairwise testing. Subsequent investigations of the reports of actual software failures showed that pairwise (two-way) testing is often useful but not sufficient [105]. In addition, test factors and test settings are always subject to various types of constraints imposed by the runtime environment and the semantics of the SUT. In Section 14.3, we discuss combinatorial (t-way) testing (CT) for $t \geq 2$ with support of constraints to exclude invalid combinations. Combinatorial testing for $t \geq 2$

became practical because of important advances in algorithms and tools for generating test suites for t-way testing with the support of constraints. A brief summary and concluding remarks appear in Section 14.4.

14.2 PAIRWISE (TWO-WAY) TESTING OF SOFTWARE SYSTEMS

The concept of OAs was formally defined by Rao [158]. OAs are mathematical arrangements of symbols that satisfy certain combinatorial properties. OAs are generalizations of combinatorial arrangements called Latin squares [157]. The matrix shown in Table 14.1 is an OA referred to as OA $(8, 2^4 \times 4^1, 2)$. The first parameter (which is 8) indicates the number of rows and the second parameter (which is $2^4 \times 4^1$) indicates that there are five columns of which four have two distinct elements each, denoted here by {0, 1}, and one column has four distinct elements, denoted here by {0, 1, 2, 3}. The third parameter (which is 2) indicates that this OA has strength 2, which means that every set of two columns contains all possible pairs of elements exactly the same number of times. Thus, every pair of the first four columns contains the four possible pairs of elements {00, 01, 10, 11} exactly twice and every pair of columns involving the fifth column contains the eight possible pairs of elements {00, 01, 02, 03, 10, 11, 12, 13} exactly once. In an OA of strength 3, every set of three columns contains all possible triplets of elements exactly the same number of times.

A fixed-value orthogonal array denoted by OA (N, v^k, t) is an $N \times k$ matrix of elements from a set of v symbols $\{0, 1, \ldots, (v-1)\}$ such that every set of t-columns contains each possible t-tuple of elements the same number of times. The positive integer t is the strength of the orthogonal array. The number of times each t-tuple appears is called its index. In the context

TABLE 14.1 Orthogonal Array OA $(8, 2^4 \times 4^1, 2)$

	1	**2**	**3**	**4**	**5**
1	0	0	0	0	0
2	1	1	1	1	0
3	0	0	1	1	1
4	1	1	0	0	1
5	0	1	0	1	2
6	1	0	1	0	2
7	0	1	1	0	3
8	1	0	0	1	3

of an OA, elements such as 0, 1, 2, . . ., $(v - 1)$ are symbols rather than numbers. The combinatorial property does not depend on the symbols that are used for the elements. A fixed-value orthogonal array may also be denoted by OA (N, k, v, t). A mixed-value orthogonal array is an extension of fixed-value OA, where $k = k_1 + \cdots + k_n$; k_1 columns have v_1 distinct elements, k_2 columns have v_2 distinct elements, . . ., and k_n columns have v_n distinct elements. The mathematics of orthogonal arrays and extensive references can be found in Hedayat et al. [85]. An electronic library of known OAs is maintained by Sloane [184].

The term *design of experiments* refers to a methodology for conducting controlled experiments in which a system is exercised (worked in action) in a purposeful (designed) manner for chosen test settings of various input variables (called test factors). The corresponding values of one or more output variables (called system responses) are measured to generate information for improving the performance of a class of similar systems. Conventional DoE methods were developed in the 1920s largely by Fisher [68,69] and his contemporaries and followers to improve agricultural production. Later, DoE methods were adapted for experiments with animals and then to improve manufacturing processes subject to uncontrolled variation. Frequently, the effects of many test factors each having multiple test settings are investigated at the same time and the DoE plans satisfy relevant combinatorial properties [22,47,93,134,185].

Genichi Taguchi promulgated a variation of conventional DoE methods in Japan (1960s to 1970s), United States (1980s), and elsewhere [190–192]. The objective in conventional DoE is to improve average response, but the objective in Taguchi DoE is to determine test settings at which the variation due to uncontrolled factors was least [88,155]. Taguchi promoted the use of OAs of strength two as templates for his DoE plans. Before Taguchi's use of OAs, they were not well known outside the world of mathematics. A salient property of DoE based on OAs is that they enable evaluation of a statistic called the *main effect* of each test factor [22]. The main effect of a test factor is the average effect over all test settings of every other factor on the objective function (which may be the average response or the average of a measure of variation of response). The evaluation of the main effects is important in DoE because for each test factor, optimal test settings should apply to the ranges of the values rather than fixed values of the other test factors.

Along with the advent of computers and telecommunication systems based on software, the problem of testing software and software-embedded systems became increasingly important in the 1980s. Taguchi inspired the use of OAs for testing software systems. Software engineers in various companies (especially Fujitsu in Japan and descendent organizations of the AT&T Bell System in the United States) started to investigate the use of DoE plans and OAs for testing software systems. The earliest papers include Sato and Shimokawa [175], Shimokawa [183], Mandl [125], Tatsumi [194], and Tatsumi et al. [195]. Tatsumi [194] is a signal paper that included two important insights. (1) In software testing, all combinations need not be tested the same number of times, only each combination needs to be tested at least once. (2) In generating test suites for testing software, invalid combinations and test cases must be excluded.

In the United States, starting in the late 1980s, Phadke [155] and his colleagues developed a tool called OATS (orthogonal array testing system) to generate test suits (based on OAs of strength 2) for Taguchi DoE and for testing software-based systems [23]. OATS was an AT&T proprietary tool for intracompany use only. The uses of OAs of strength 2 assured that all test settings for each test factor and all pairs of test settings for every pair of test factors were tested (executed).

Thus began pairwise (two-way) testing of software systems. Pairwise testing is a type of dynamic testing in which the SUT is exercised for a test suite that satisfies the property that for every pair of test factors all possible pairs of the discrete test settings are tested (at least once). Pairwise testing is an economical alternative to exhaustive testing of all combinations of the test factors. For example, if we had nine test factors with the combinatorial test structure $3^3 4^4 5^2$, then exhaustive testing would require $3^3 4^4 5^2 = 172{,}800$ tests. However, pairwise testing requires only 29 test cases, a dramatically smaller number than 172,800 test cases required for exhaustive testing. Thus, pairwise testing could greatly improve the efficiency of dynamic testing of software. Many faults in software involve only one or two factors, so, pairwise testing was found to be effective in detecting faults.

When an AT&T test engineer [179] tried to use OATS to specify "client test configurations for a local area network product," he realized the limitations of OATS and the limitations of using OAs to construct test suits for software testing. Often, an OA matching the required combinatorial test structure does not exist. Also, frequently, OA-based test suites included invalid test cases. Suppose four factors have two test settings each and one has three test settings. Thus, the test structure is $2^4 \times 3^1$. An OA of strength 2 matching the test structure $2^4 \times 3^1$ does not exist. (It is mathematically

TABLE 14.2 Combinatorial Arrangement for the Test Structure $2^4 \times 3^1$

	1	**2**	**3**	**4**	**5**
1	0	0	0	0	0
2	1	1	1	1	0
3	0	0	1	1	1
4	1	1	0	0	1
5	0	1	0	1	2
6	1	0	1	0	2
7	0	1	1	0	2
8	1	0	0	1	2

impossible.) In such cases, a suitable OA is modified to fit the need. For example, if the elements in the row 7 and the row 8 of the column 5 of OA (8, $2^4 \times 4^1$, 2) shown in Table 14.1 are changed from 3 to 2, we get the combinatorial arrangement shown in Table 14.2. Table 14.2 is not an OA but it covers all pairs of test settings and it can be used to construct a pairwise testing suite for the test structure $2^4 \times 3^1$.

Suppose the five test factors of combinatorial test structure $2^4 \times 3^1$ were (1) operating system (OS) with two test settings {XP, Linux}, (2) browser with two test settings {Internet Explorer (IE), Firefox}, (3) protocol with two test settings {IPv4, IPv6}, (4) CPU type with two test settings {Intel, AMD}, and (5) database management system (DBMS) with three test settings {MySQL, Sybase, Oracle}. Then, a test suite for pairwise testing based on Table 14.2 is shown in Table 14.3. To obtain Table 14.3 from Table 14.2, the five test factors OS, Browser, Protocol, CPU type, and the DBMS are associated with the five columns of Table 14.2 and the elements in the columns {0, 1, 2} are replaced with the test settings of the respective test factors. The 8 rows of Table 14.3 form a test suite for pairwise testing.

TABLE 14.3 Pairwise Testing Suite Based on Table 14.2

Tests	**OS**	**Browser**	**Protocol**	**CPU**	**DBMS**
1	XP	IE	IPv4	Intel	MySQL
2	Linux	Firefox	IPv6	AMD	MySQL
3	XP	IE	IPv6	AMD	Sybase
4	Linux	Firefox	IPv4	Intel	Sybase
5	XP	Firefox	IPv4	AMD	Oracle
6	Linux	IE	IPv6	Intel	Oracle
7	XP	Firefox	IPv6	Intel	Oracle
8	Linux	IE	IPv4	AMD	Oracle

Since the browser IE does not run on the OS Linux, the pair {Linux, IE} appearing in test cases 6 and 8 is invalid. Therefore, these two test cases are not legitimate and cannot be executed. In that case, other valid pairs of test settings covered by these two test cases are not tested, for example, the pairs {IPv6, Oracle}, {Intel, Oracle}, {IPv4, Oracle}, and {AMD, Oracle} are not tested. Thus, the test suite shown in Table 14.3 with test cases 6 and 8 omitted would not test all valid pairs of test settings.

Since OAs are not available for many combinatorial test structures and since test suites based on OAs may include invalid test cases, in the early 1990s, Sherwood [179] developed a tool called CATS (constrained array test system) to generate test suites that covered all valid combinations of test settings with a small number of test cases. The test suite generation tool CATS (like OATS) was an AT&T proprietary tool for intracompany use only. Test suites generated by CATS were related to mathematical arrangements of symbols called covering arrays. The concept of covering arrays (CAs) was formally defined by Sloane [182]. Significant earlier contributions leading to the concept of CAs include Renyi [159], Kleitman and Spencer [94], and Roux [160]. Additional references on CAs can be found in Lawrence et al. [114] and Torres-Jimenez et al. [198].

A fixed-value covering array denoted by CA (N, v^k, t) is an $N \times k$ matrix of elements from a set of v symbols {0, 1, …, (v – 1)} such that every set of t-columns contains each possible t-tuple of elements at least once. The positive integer t is the strength of the covering array. A fixed value orthogonal array may also be denoted by CA (N, k, v, t). A mixed-value covering array is a generalization of the concept of fixed value CA, where $k = k_1 + k_2 + \cdots + k_n$; k_1 columns have v_1 distinct elements, k_2 columns have v_2 distinct elements, …, and k_n columns have v_n distinct elements. Unlike an OA, a CA need not be balanced in the sense that not all t-tuples need to appear the same number of times. All OAs are CAs but not all CAs are OAs. Thus, the concept of covering arrays is a generalization of OAs. (A fixed-value orthogonal array of index one is the smallest covering array.)

Methods for constructing CAs can be put into three categories: (1) *algebraic methods* (e.g., Bush [35], Roux [160], Sloane [182], Hartman [83], and others); (2) *metaheuristic methods* such as simulated annealing and tabu search (e.g., Cohen et al. [49], Colbourn [53], Nurmela [146], Torres-Jimenez et al. [198], and others); and (3) *greedy search methods* (e.g., Lei and Tai [119], Tai and Lei [193], Lei et al. [116], Forbes [71], and others). Algebraic methods apply only to certain special combinatorial test structures; however, when they apply, they are extremely fast mathematical techniques and may produce CAs of smallest possible size. Metaheuristic methods are computationally intensive and they have produced some CAs of the smallest size known. Greedy methods are faster than metaheuristic methods; they apply to arbitrary test structures but may or may not produce smallest-size CAs. A particular greedy algorithm has produced some CAs of the smallest size known [71]. The three approaches are sometimes combined to yield additional methods for constructing covering arrays [116,180,205], and others.

Colbourn maintains a web page of smallest known sizes (N) of various covering arrays CA (N, k, v, t) of strength t up to seven [51]. The site gives the current best-known upper bound on the least number of rows (the least number of rows is called the covering array number). Web pages of CA tables include the following: Forbes [72], Nurmela [147], and Torres-Jimenez [199].

The use of covering arrays for software testing was promulgated by Dalal and Mallows [64] among others. They noted that evaluation of main effects is important in DoE, and OAs enable evaluations of the main effects; however, in testing software systems, there is no need to evaluate the main effects of test factors. Instead, interest lies in covering all pairs (in general all t-tuples) of test settings. Therefore, covering arrays are better suited than OAs for testing software. It turns out that CAs have several advantages over OAs. (1) CAs can be constructed for test factors with arbitrary combinatorial test structures of unequal numbers of test settings. (2) For a combinatorial test structure, if an OA exists, then a CA of the same or fewer test runs can be obtained. (3) CAs can be constructed for any required strength (t-way) testing, while OAs are generally limited to strength 2 and 3 [184]. (4) In generating test suites for combinatorial testing, certain combinations may need to be excluded (e.g., invalid combinations); these combinations can be deliberately excluded from test suites based on CAs.

To our knowledge, the first publicly available tool for generating test suites based on CAs for pairwise (and higher strength) testing of software systems was AETG [50,51]. In 1998, a graduate student Yu Lei developed an algorithm called in parameter order (IPO) for generating test suites for pairwise testing based on CAs, which excluded invalid pairs of test settings [117,193]. The usefulness of CAs for pairwise testing led to great interest among mathematicians and computer scientists to develop tools for generating test suites based on CAs.

14.3 COMBINATORIAL *t*-WAY TESTING OF SOFTWARE SYSTEMS

A team of NIST researchers investigated 15 years of recall data due to failures of software embedded in medical devices [212] and did a series of follow-up studies, including failure reports for a browser, a server, and a NASA database system [103–105]. The primary purpose of the initial investigation was to generate insights into the kinds of software testing that could have detected the underlying faults and prevented the failures in use. In the subsequent investigations, the researchers were able to determine the numbers of individual factors that were involved in the faults underlying actual failures of software, leading to the empirically derived interaction rule introduced in Chapter 1: *Most failures are induced by single factor faults or by the joint combinatorial effect (interaction) of two factors, with progressively fewer failures induced by interactions between three or more factors.* The maximum degree of interaction in actual faults so far observed is six.

Combinatorial (*t*-way) testing may be regarded as an adaptation of the DoE methods for testing software systems because in both cases, information about a system is gained by exercising it and the test suite (DoE plan) satisfies relevant combinatorial properties. Unlike DoE, in CT for each possible test case, the expected behavior of the system is predetermined from the available information such as a mathematical model of the SUT. A software tool is often used to check whether the actual behavior matches the expected behavior and to make a verdict of passing or failing for each test case. CT can be used in a broad range of test problems, as discussed in other chapters. These include (1) test database tables and (2) test inputs according to state models [181], as well as (3) concurrent systems [118], (4) web applications [207], (5) security testing of access control implementations [96], (6) navigation of dynamic web structures [207], (7) optimization of discrete event simulation models, (8) analyzing system state-space coverage [130], (9) detecting deadlocks for varying network configurations [107], (10) detecting buffer overflow vulnerabilities [209], (11) conformance testing for standards [133], (12) event sequence testing [110], and (13) prioritizing user session-based test suites for web applications [173]. Generally, CT does not require access to the source code; however, once one or more failure-inducing combinations have been identified, the source code may be needed in the follow-up investigations to reveal the underlying faults in the SUT.

A failure inducing fault may not be detected when the test factors and settings associated with that fault are not included in the test plan and exercised. When continuous-valued factors are involved, their chosen discrete test settings preclude testing certain values. Therefore, testing can detect faults but it cannot guarantee their absence. It is difficult to categorize the kinds of faults a software system can have. Multiple testing methods during and after development are generally needed to assure correctness of software. Combinatorial testing complements other software testing methods.

14.4 CHAPTER SUMMARY

Combinatorial (t-way) testing may be regarded as an adaptation of design of experiment methods for testing software systems because in both cases, information about the SUT is gained by exercising it and the test plan satisfies relevant combinatorial properties. Indeed, CT evolved from the use of DoE plans based on OAs for generating test suits for software testing. CT requires specification of test factors and discrete test settings for each. A test case is a combination of one selected test setting for each test factor. CT began as pairwise (two-way) testing in which the SUT is exercised for a test suite of test cases, which satisfies the property that for every pair of test factors all possible pairs of the test settings are tested at least once. First, orthogonal arrays were used as templates for constructing pairwise test suites. However, OAs could not support constraints among the test settings of test factors. Therefore, covering arrays were found to be better suited than OAs for combinatorial t-way testing for $t \geq 2$. Investigations of actual faults indicated that while pairwise ($t = 2$) testing is useful, it usually is not adequate for strong assurance. Therefore, combinatorial t-way testing for t greater than 2 may be needed. Combinatorial t-way testing for $t > 2$ is now possible because efficient and free downloadable tools for generating test suites for t-way testing with support of constraints (to exclude invalid combinations) have become available. As with any test method, CT can detect faults but it cannot guarantee their absence. CT is one of many complementary methods for software assurance.

CHAPTER 15

Algorithms for Covering Array Construction

Linbin Yu and Yu Lei

THIS CHAPTER INTRODUCES SOME general strategies for covering array construction, and explains two widely used algorithms, namely, AETG and IPOG. We also present several practical considerations in covering array construction, including constraint handling, mixed-strength covering array, and covering array extension.

15.1 OVERVIEW

Existing approaches to covering array construction can be classified as either computational or algebraic [84,225]. Computational approaches involve explicitly enumerating and covering every t-way combination, whereas algebraic approaches construct test sets based on some predefined rules without explicit enumeration and covering of combinations.

15.1.1 Computational Approaches

Cohen et al. proposed a strategy, called AETG (automatic efficient test generator), which builds a test set "one-test-at-a-time" until all the t-way combinations are covered [50,61]. A greedy algorithm is used to construct the tests such that each subsequent test covers as many uncovered t-way combinations as possible. Several variants of this strategy have been reported that use slightly different heuristics in the greedy construction process [24,226]. The AETG strategy and its variants are later generalized into a general framework [25].

Lei and Tai developed a pairwise testing strategy, called IPO (in-parameter-order), which builds a pairwise test set for the first two parameters, extends the test set to cover the first three parameters, and continues to extend the test set until it builds a pairwise test set for all the parameters [119,193]. This strategy is later extended to general t-way testing, and the extended strategy is referred to as IPOG (in-parameter-order-general) [117].

More recently, techniques such as hill climbing and simulated annealing have been applied to covering array construction [225]. These techniques start from a preexisting test set and then apply a series of transformations until a test set is reached that covers all the combinations that need to be covered, preferably in as few tests as possible. Techniques like hill climbing and simulated annealing can produce smaller test sets than AETG and IPOG, but they typically take longer to complete. Novel methods based on formal grammars have also been developed [168].

The main advantage of computational approaches is that they can be applied to an arbitrary system configuration. That is, there is no restriction on the number of parameters and the number of values each parameter can take in a system configuration. Moreover, it is relatively easy to adapt computational approaches for test prioritization [26] and constraint handling [77]. However, computational approaches involve explicitly enumerating all possible combinations to be covered. When the number of combinations is large, explicit enumeration can be prohibitive in terms of both the space for storing these combinations and the time needed to enumerate them. In addition, computational approaches are typically greedy, that is, they construct tests in a locally optimized manner, which does not necessarily lead to a globally optimized test set. Thus, the size of a test set generated may not be minimal.

15.1.2 Algebraic Approaches

In algebraic approaches, test sets are derived from covering arrays that are constructed by predefined rules without requiring any explicit enumeration of combinations. There are two main types of algebraic approaches for constructing covering arrays. In the first type of approach, a covering array is constructed directly by computing a mathematical function for the value of each cell based on its row and column indices. These approaches are generally extensions of the mathematical methods for constructing orthogonal arrays [35,125]. The second type of algebraic approach is based on the idea of recursive construction, which allows larger covering arrays to be constructed from smaller covering arrays [219,227]. For example, the

D-construction approach uses a pair of 2-way and 3-way covering arrays of k columns to construct a 3-way covering array of $2k$ columns [228].

Since algebraic approaches do not enumerate any combinations, they are immune from the combinatorial explosion problem. The computations involved in algebraic construction are usually lightweight. Thus, algebraic approaches can be extremely fast. Unfortunately, algebraic approaches often impose serious restrictions on the system configurations to which they can be applied. For example, many approaches for constructing orthogonal arrays require that the domain size be a prime number or a power of a prime number. This significantly limits the applicability of algebraic approaches for software testing. Finally, test prioritization [26] and constraint handling [77] can be more difficult for algebraic approaches.

15.2 ALGORITHM AETG

As mentioned earlier, the AETG algorithm builds a t-way test set "one-test-at-a-time" until all the t-way combinations are covered. Cohen et al. [61] showed that the number of tests in a t-way test set grows logarithmically in the number of parameters if an optimal test, in terms of covering the maximum number of uncovered combinations, can be selected at each step. The key challenge of the AETG algorithm is to find such an optimal test at each step. This is accomplished by choosing a test that covers the most previously uncovered combinations from a set of randomly generated tests.

Figure 15.1 shows the pseudocode for the AETG algorithm. It takes three arguments: (1) an integer t specifying the test strength; (2) a parameter set ps containing the input parameters and their values; (3) an integer m specifying the number of candidate tests to be generated at each step. The output of this algorithm is a t-way test set (or covering array).

Assume that a system contains k parameters, and the i-th parameter has d_i different values. The AETG algorithm will first find all uncovered combinations (line 2). Then, it generates one test at time, until all combinations are covered (line 3). At each iteration, the AETG algorithm will first generate m different candidate tests randomly (lines 5–12). The set of candidate tests is generated as follows. For simplicity, we assume the test strength is 2. It first chooses a parameter p and then assigns a value v by selecting a value that appears the most times in the uncovered combinations, and reorder all others parameters randomly (line 7). Denote the reordered parameters as $p_1, p_2, \ldots, p_k$. Then, it selects a value for each parameter in their new order (lines 8–11). Assume that we have selected

```
algorithm AETG (int t, ParameterSet ps, int m)
{
1. initialize test set ts to be an empty set
2. let π be the set of all t-way combinations to be covered
3. while (π is not empty) {
4.   let p.v be the parameter value that appears the most number of times in π
5.   //generate m candidates
6.   for (int i = 0; i < m; i ++) {
7.     let p be the first parameter, and reorder all others parameters randomly.
         denote the reordered parameters as p1, p2, ... pk.
8.     //select values for each candidate
9.     for (int j = 1; j <= k; j ++) {
10.       select a value for the j-th parameter pj such that the most uncovered
            combinations can be covered
11.     } //end for in line 9
12.  }//end for in line 6
13.  select a test τ from m candidates such that it covers the most uncovered
       combinations
14.  add τ into ts and remove from π the set of combinations covered by τ
15.} //end while in line 3
16. return ts
}
```

FIGURE 15.1 Algorithm AETG.

values for the first j parameters $p_1, p_2, \ldots, p_j$. Now, choose a value v_{j+1} for p_{j+1} such that the first $j + 1$ parameters contain the greatest number of new pairs (line 10). After all candidate tests are generated, it selects the best one that can cover the most uncovered combinations (line 13). The covered combinations will be removed. The algorithm terminates when all combinations are covered (line 15).

The AETG algorithm is nondeterministic. Multiple test sets can be generated using different random seeds. Thus, one may run the AETG algorithm for multiple times and use the smallest test set.

In the following, we use an example to illustrate how the AETG algorithm works. Assume that a system contains four parameters, denoted as p_1, p_2, p_3, and p_4, each of which has two values {0, 1}. Also assume that the test strength t is 2, and the number m of candidate tests is 3.

The AETG algorithm first computes the set of all possible 2-way combinations or pairs, which is shown in Figure 15.2a. There are a total of 24 2-way combinations. Next, it builds a 2-way test set. Assume the first two tests are {0,0,0,0} and {1,1,1,1}. Next, it generates the third test as follows:

We choose p_1 as the first parameter and 0 as its value since it appears the most times (3) in the uncovered pairs. Now, we generate three candidate tests randomly. For each candidate test, we first reorder the parameters p_2,

(a)

p_1	p_2	p_1	p_3	p_1	p_4	p_2	p_3	p_2	p_4	p_3	p_4
0	0	0	0	0	0	0	0	0	0	0	0
0	1	0	1	0	1	0	1	0	1	0	1
1	0	1	0	1	0	1	0	1	0	1	0
1	1	1	1	1	1	1	1	1	1	1	1

(b)

	p_1	p_2	p_3	p_4	**Newly covered pairs**
Step 1	0	0	0	0	6
Step 2	1	1	1	1	6
Step 3	0	0	1	1	4
Step 4	1	1	0	0	4
Step 5	0	1	1	0	2
Step 6	1	0	0	1	2

(c)

Candidate test 1	Reordered parameters	p_1	p_4	p_2	p_3	Covered pairs
	Selected values	0	1	0	1	3
Candidate test 2	Reordered parameters	p_1	p_3	p_2	p_4	Covered pairs
	Selected values	0	1	0	1	3
Candidate test 3	Reordered parameters	p_1	p_2	p_4	p_3	Covered pairs
	Selected values	0	1	0	1	3

FIGURE 15.2 Example of test generation using the AETG algorithm. (a) All 24 uncovered pairs; (b) generated tests; (c) candidate tests in step 3.

p_3, and p_4 randomly, and then select values for these parameters in order. Assume the orders of parameters in three candidate tests are $\{p_1, p_4, p_2, p_3\}$, $\{p_1, p_3, p_2, p_4\}$, and $\{p_1, p_2, p_4, p_3\}$, respectively. As shown in Figure 15.2c, we select values for these parameters. The values are selected such that the greatest number of new pairs can be covered. Take the first candidate as an example, the value for p_1 fixed to 0. The value for the second parameter p_4 is 1, as $\{p_1 = 0, p_4 = 1\}$ covers 1 new pair and $\{p_1 = 0, p_4 = 0\}$ does not. Similarly, we select value 0 for the third parameter p_2 and value 1 for the last parameter p_3. This candidate test $\{p_1 = 0, p_4 = 1, p_2 = 0, p_3 = 1\}$ covers four new pairs. We fix the other two candidate tests in the same way. All these three candidate tests can cover four new pairs, and we pick up the first one as the third test {0, 0, 1, 1}.

The remaining tests are generated in the same way until all uncovered pairs are covered.

15.3 ALGORITHM IPOG

The IPOG algorithm generalizes the IPO strategy from pairwise testing to general t-way testing. Compared to the AETG algorithm, the IPOG algorithm is deterministic, that is, it produces the same test set for the same system configuration. Covering one parameter at a time also allows IPOG to achieve a lower order of complexity than AETG [193].

Figure 15.3 shows the IPOG algorithm. It takes as input two arguments: (1) an integer t specifying the test strength and (2) a parameter set *ps* containing the input parameters (and their values). The output of this algorithm is a t-way test set (covering array) for the parameters in *ps*. The number of parameters in *ps* is assumed to be no less than t.

The IPOG algorithm begins by initializing test set *ts* to be empty (line 1) and sorting the input parameters in a nonincreasing order of their domain sizes (line 2). Test set *ts* will be used to hold the resulting test set. Next, the algorithm builds a t-way test set for the first t parameters. This is done by adding to set *ts* a test for every combination of the first t parameters (line 3).

If the number of parameters k is greater than the test strength t, the remaining parameters are covered, one at each iteration, by the outermost

```
Algorithm IPOG-Test (int t, ParameterSet ps)
{
1. initialize test set ts to be an empty set
2. sort the parameters in set ps in a non-increasing order of their domain sizes, and denote
      them as P1, P2, ..., and Pk
3. add into test set ts a test for each combination of values of the first t parameters
4. for (int i = t + 1; i ≤ k; i ++){
5.    let π be the set of all t-way combinations of values involving parameter Pi and any
         group of (t-1) parameters among the first i-1 parameters
6.    // horizontal extension for parameter Pi
7.    for (each test τ= (v1, v2, ..., vi-1) in test set ts) {
8.       choose a value vi of Pi and replace τ with τ' = (v1, v2, ..., vi-1, vi) so that τ' covers
            the most number of combinations of values in π
9.       remove from π the combinations of values covered by τ'
10.   } // end for at line 7
11.   // vertical extension for parameter Pi
12.   for (each combination σ in set π){
13.      if (there exists a test τ in test set ts such that it can be changed to cover σ) {
14.         change test τ to cover σ
15.      } else {
16.         add a new test to cover σ
17.      } // end if at line 13
18.   } // end for at line 12
19. }// end for at line 4
20. return ts;
}
```

FIGURE 15.3 Algorithm IPOG.

for loop (line 4). Covering parameter P_i means that test set *ts* is extended to become a t-way test set for parameters $\{P_1, P_2, \ldots, P_i\}$. Let P_i be the parameter that the current iteration is trying to cover. The IPOG algorithm first computes the set π of combinations that must be covered to cover parameter P_i (line 5). Note that test set *ts* is already a t-way test set for parameters $\{P_1, P_2, \ldots, P_{i-1}\}$. To build a t-way test set for $\{P_1, P_2, \ldots, P_i\}$, it is sufficient to cover all the t-way combinations involving P_i and any group of $(t - 1)$ parameters among $P_1, \ldots,$ and P_{i-1}, which are the parameters that have already been covered.

The combinations in set π are covered in the following two steps:

- *Horizontal growth*: This step adds a value of parameter P_i to each of the existing tests already in test set *ts* (lines 7–10). These values are chosen in a greedy manner. That is, at each step, the value chosen is a value that covers the most combinations in set π (line 8). A tie needs to be broken consistently to ensure that the resulting test set is deterministic. After a value is added, the newly covered combinations are removed from set π (line 9).
- *Vertical growth*: At this step, the remaining combinations, that is, combinations that are not covered by horizontal growth, are covered, one at a time, either by changing an existing test (line 14) or by adding a new test (line 16). Let σ be a t-way combination that involves parameters $P_{k1}, P_{k2}, \ldots,$ and P_{kt}. An existing test τ can be changed to cover σ if and only if the value of P_{ki}, $1 \le i \le t$, in τ is either the same as in σ or a *don't care* value. A *don't care* value is a value that can be replaced by any value without affecting the coverage of a test set. A don't care value may be introduced if no existing test can be changed to cover σ. In this case, a new test needs to be added in which the value of P_{ki}, $1 \le i \le t$, is assigned the same value as in σ, and the other parameters are assigned *don't care* values.

In the following, we illustrate the IPOG algorithm using the same example system in the previous section. Recall that the example system contains four parameters p_1, p_2, p_3, and p_4, and each parameter has two available values {0, 1}. The test strength is 2. We also select the first candidate value if there is a tie.

In the first step, we sort all parameters according to their domain sizes. Since all these four parameters have the same domain, we skip this step in this example. Then, the IPOG algorithm generates all the combinations for the first two parameters p_1 and p_2. Now, we obtain the 2-way covering array for the first two parameters, as shown in Figure 15.4a.

Now, we extend the current test set for the third parameter p_3. In the horizontal growth, we append values for each test such that the most uncovered pairs can be covered. We select 0 and 1 for test 1 and test 2, respectively, as they can cover the most (2) uncovered pairs. For test 1, it covers $\{p_1 = 0, p_3 = 0\}$ and $\{p_2 = 0, p_3 = 0\}$; for test 2, it covers $\{p_1 = 1, p_3 = 1\}$ and $\{p_2 = 1, p_3 = 1\}$. Similarly, we select 1 for test 3, since it covers $\{p_1 = 1, p_3 = 1\}$ and $\{p_2 = 0, p_3 = 1\}$, and select 0 for test 4, since it covers $\{p_1 = 1, p_3 = 0\}$ and $\{p_2 = 1, p_3 = 0\}$. Since there are no uncovered pairs for the first three parameters, the vertical growth is not necessary. Now, we obtain the 2-way covering array for the first three parameters, as shown in Figure 15.4b.

Similarly, we apply horizontal growth for the last parameter p_4, as shown in Figure 15.4c. Since there are three uncovered pairs $\{p_1 = 1, p_4 = 1\}$, $\{p_2 = 0, p_4 = 1\}$, and $\{p_3 = 0, p_4 = 1\}$, we apply vertical growth for p_4. During the vertical growth, we first add an empty fifth test {*, *, *, *}, in which "*" denotes the *don't care* value. Then, we find the first missing tuple $\{p_1 = 1, p_4 = 1\}$, and change test 5 to {1, *, *, 1} to cover it. Next, we find the second missing tuple $\{p_2 = 0, p_4 = 1\}$. Since test 5 is compatible with this missing tuple, we change test 5 to {1, 0, *, 1} to cover the second missing tuple. Similarly, we change test 5 to {1, 0, 0, 1} to cover the last missing tuple. Since the new test covers all three uncovered pairs, the vertical growth is finished, as shown in Figure 15.4d. Now the algorithm terminates and we have generated a 2-way test set for all four parameters.

(a)

	p1	p2
test 1	0	0
test 2	0	1
test 3	1	0
test 4	1	1

(b)

	p1	p2	p3
test 1	0	0	**0**
test 2	0	1	**1**
test 3	1	0	**1**
test 4	1	1	**0**

(c)

	p1	p2	p3	p4
test 1	0	0	0	**0**
test 2	0	1	1	**1**
test 3	1	0	1	**0**
test 4	1	1	0	**0**

(d)

	p1	p2	p3	p4
test 1	0	0	0	0
test 2	0	1	1	1
test 3	1	0	1	0
test 4	1	1	0	0
test 5	**1**	**0**	**0**	**1**

FIGURE 15.4 Example of test generation using the IPOG algorithm. (a) The initial matrix; (b) the horizontal growth for p_3; (c) the horizontal growth for p_4; (d) the vertical growth for p_4.

15.4 COST AND PRACTICAL CONSIDERATIONS

We discuss several practical considerations for covering array construction, including constraint handling, mixed strength, and covering array extension.

15.4.1 Constraint Handling

In real-life applications, the user often needs to prohibit certain combinations of parameter values from appearing in a test. For example, the user may want to ensure that a web page can be displayed correctly in different web browsers running on different operating systems. If the web browser is IE, the operating system cannot be MacOS. Therefore, no test should contain the combination of {IE, MacOS}.

In general, there are two types of constraints, environment constraints and system constraints. Environment constraints are imposed by the runtime environment of the system under test (SUT). In the example introduced earlier, when we test a web application to ensure that it works in different operating systems and browsers, a combination of Linux and IE does not occur in practice and cannot be tested. In general, combinations that violate environment constraints could never occur at runtime and thus must be excluded from a test set.

System constraints are imposed by the semantics of the SUT. For example, a flight reservation system may impose a constraint that the number of passengers on a flight must be no more than the number of seats available on the airplane assigned for the flight. Invalid combinations that do not satisfy system constraints may still be rendered to the SUT at runtime. When this occurs, these combinations should be properly rejected by the SUT. Therefore, it is important to test these combinations for the purpose of robustness testing, that is, making sure that the SUT is robust when invalid combinations are presented. Note that to avoid potential mask effects, robustness testing often requires that each test contains only one invalid combination.

There are two general approaches to handling constraints with covering array generators. The first is to transform the input model without changing the test generation algorithm. For example, assume that there exists a constraint, $a > b$, between two parameters a and b. A new input model can be created by replacing these two parameters with a new parameter c whose domain consists of all the combinations of parameters a and b that satisfy this constraint. A test generation algorithm that does not support constraints can be applied to this new model to create a combinatorial test set. The second approach is to modify the test generation algorithm such that constraints are handled properly during the actual test generation process. For example, many algorithms are greedy algorithms that typically create a test by choosing, from a pool of candidates, one that covers the most number of uncovered tests. Such algorithms can be modified such that they require each candidate in the pool satisfy all the constraints. Compared to the first approach, this approach often produces a smaller test set, but at the cost of more execution time.

In a greedy approach [63], constraints are represented by a set of logical expressions, and then are converted to a list of forbidden tuples. Later in the generation stage, valid tests are generated using an AETG-like greedy approach. The list of forbidden tuples is used to ensure that all generated tests are valid. Another approach [229] adopts a different greedy generation algorithm. The constraints are provided by the user in the form of forbidden tuples. Constraints can also be handled using constraint solving. A Boolean satisfiability problem (SAT) solver can be integrated into the simulated annealing algorithm and the AETG-like generation algorithm [50]. Several recent works focus on a formal logical strategy for test generation with constraints [230–232]. In these approaches, test generation is modeled as a formal logic problem that can be solved by existing SAT modulo theory (SMT) solvers. Constraints over value combinations can be naturally described in the system model, and can be automatically handled by an SMT solver.

15.4.2 Mixed-Strength Covering Arrays

A covering array is usually constructed using a single strength. That is, a single strength t is applied to all parameters. However, in many systems, some parameters may interact with each other more than other parameters. A mixed-strength covering array allows different groups of parameters to be tested with different strengths.

Algorithms that build covering arrays of a single strength can be modified to build covering arrays of mixed strengths. Conceptually speaking, there are two major steps in the construction of a covering array. The first step is to compute the target set of t-way combinations, that is, all the t-way combinations that need to be covered. The second step is to generate tests to cover every t-way combination in the target set. To support the construction of mixed-strength covering arrays, the first step needs to be modified such that the target set includes all the t-way combinations of different strengths. In contrast, the second step largely remains the same.

(a)

p1	p4
0	0
0	1
1	0
1	1

p2	p4
0	0
0	1
1	0
1	1

p3	p4
0	0
0	1
1	0
1	1

p1	p2	p3
0	0	0
0	0	1
0	1	0
0	1	1
1	0	0
1	0	1
1	1	0
1	1	1

Strength 2 for {p1, p2, p3, p4}
Strength 3 for {p1, p2, p3}

(b)

	p1	p2	p3	p4
test 1	0	0	0	0
test 2	0	0	1	1
test 3	0	1	0	0
test 4	0	1	1	0
test 5	1	0	0	1
test 6	1	0	1	0
test 7	1	1	0	1
test 8	1	1	1	0

FIGURE 15.5 Example of mixed-strength covering array. (a) All combinations need to cover; (b) the generated test set.

That is, tests can still be generated in the typical manner, that is, each new test is generated such that it covers as many uncovered t-way combinations as possible.

We use the same system in the previous section as an example. The system contains four parameters p_1, p_2, p_3, and p_4. Each parameter has two available values {0, 1}. The test strength for all parameters is 2. We also have test strength 3 for $\{p_1, p_2, p_3\}$. The combinations that need to be covered are shown in Figure 15.5a and the covering array with mixed strength is shown in Figure 15.5b. One may verify that all eight value combinations for $\{p_1, p_2, p_3\}$ are covered.

15.4.3 Extension of an Existing Test Set

There are two cases in which a user may want to extend an existing test set. In the first case, a system may be already tested with a test set that is not t-way complete and the user wants to augment the test set to achieve t-way coverage. In the second case, the configuration of a system may change as the system evolves. For example, a new parameter or parameter value may be introduced. In this case, an existing test set that covers all the t-way combinations of the old configuration will no longer cover all the t-way combinations of the new configuration. Such a test set can be extended to achieve t-way coverage for the new configuration. Doing so allows one to reuse the effort that has already been spent with the existing test set. The number of the new tests added into the existing test set could be significantly smaller than the number of tests if a t-way test set is generated from scratch for the new configuration.

Algorithms that build covering arrays from scratch can be modified to extend an existing test set. As mentioned earlier, construction of a covering array consists of two major steps, that is, computing the target set and generating tests to cover the target set. Similar to support for mixed strengths, the first step needs to be modified such that combinations that are already covered by the existing test set are excluded from the target set. The second step largely remains the same. That is, tests can still be generated in the typical manner, that is, each new test is generated such that it covers as many uncovered t-way combinations as possible.

15.5 CHAPTER SUMMARY

Many algorithms have been developed for efficient construction of covering arrays. These algorithms can be classified largely into computational and algebraic methods. Computational methods are used more widely than algebraic methods because they do not impose any restriction on the subject systems and can be easily adapted to include practical features such as constraint support, mixed strengths, and extension from an existing test set.

Two algorithms, AETG and IPOG, are among the most commonly used in practice. Both algorithms adopt a greedy framework, but they differ in that AETG builds a complete test at a time, whereas IPOG covers one parameter at a time. A number of variants of the two algorithms have been reported. These algorithms employ different heuristics to optimize the test generation process. These algorithms have their own advantages and disadvantages. Some algorithms build a smaller test set, whereas other algorithms can be faster.

Support for constraint handling, mixed strength, and extension of an existing test set is important for practical applications. Constraint handling allows one to exclude combinations that are not valid based on domain semantics. Mixed strength allows one to test parameters that interact closely with each other at a higher strength than other parameters. Extension of an existing test allows one to reuse test effort. Computational methods such as AETG and IPOG can be easily adapted to support these practical features.

REVIEW

In the following questions, consider a system S consisting of three parameters, P_1, P_2, and P_3, where P_1 has two values 0 and 1, and P_2 and P_3 have three values 0, 1, and 2.

Q: Determine the set of all possible pairs that must be covered by a 2-way test set.

A:

p_1	p_2	p_1	p_3	p_2	p_3
0	0	0	0	0	0
0	1	0	1	0	1
0	2	0	2	0	2
1	0	1	0	1	0
1	1	1	1	1	1
1	2	1	2	1	2
				2	0
				2	1
				2	2

Q: Assume that a 2-way test set has been constructed that covers P_1 and P_2. Determine the set of all possible pairs that must be covered to extend this test set to cover P_3.

A:

p_1	p_3	p_2	p_3
0	0	0	0
0	1	0	1
0	2	0	2
1	0	1	0
1	1	1	1
1	2	1	2
		2	0
		2	1
		2	2

Q: Use the AETG algorithm to construct a 2-way test set for system S. Show the intermediate steps.

A:

1. Fix $P_3 = 0$. Reorder P_1, P_2 and generate the first test: $\{P_1 = 0, P_2 = 0, P_3 = 0\}$
2. Fix $P_3 = 1$. Reorder P_1, P_2 and generate a test: $\{P_1 = 1, P_2 = 1, P_3 = 1\}$
3. Fix $P_3 = 2$. Reorder P_1, P_2 and generate a test: $\{P_1 = 1, P_2 = 0, P_3 = 2\}$
4. Fix $P_1 = 0$. Reorder P_2, P_3 and generate a test: $\{P_1 = 0, P_2 = 1, P_3 = 2\}$
5. Fix $P_2 = 2$. Reorder P_1, P_3 and generate a test: $\{P_1 = 1, P_2 = 2, P_3 = 0\}$

6. Fix $P_2 = 2$. Reorder P_1, P_3 and generate a test: $\{P_1 = 0, P_2 = 2, P_3 = 1\}$
7. Fix $P_2 = 0$. Reorder P_1, P_3 and generate a test: $\{P_1 = 0, P_2 = 0, P_3 = 1\}$
8. Fix $P_2 = 0$. Reorder P_1, P_3 and generate a test: $\{P_1 = 0, P_2 = 1, P_3 = 0\}$
9. Fix $P_2 = 0$. Reorder P_1, P_3 and generate a test: $\{P_1 = 0, P_2 = 2, P_3 = 2\}$

Q: Use the IPOG algorithm to construct a 2-way covering array for system S. The three parameters should be covered in the given order, that is, P_1, P_2, and P_3. Show the intermediate steps.

A:

1. Generate six tests for P_1 and P_2.

p_1	p_2
0	0
0	1
0	2
1	0
1	1
1	2

2. Apply horizontal extension to each test.

p_1	p_2	p_3
0	0	0
0	1	1
0	2	2
1	0	1
1	1	2
1	2	0

3. Add three tests to cover the remaining uncovered pairs.

p_1	p_2	p_3
0	0	0
0	1	1
0	2	2
1	0	1
1	1	2
1	2	0
*	2	0
*	1	0
*	2	1

Appendix A: Mathematics Review

This appendix reviews a few basic facts of combinatorics and regular expressions that are necessary to understand the concepts in this publication. Introductory treatments of these topics can be found in textbooks on discrete mathematics.

A.1 COMBINATORICS

A.1.1 Permutations and Combinations

For n variables, there are $n!$ permutations and $\binom{n}{t} = n!/\, t!(n-t)!$ ("n choose t") combinations of t variables, also written for convenience as $C(n, t)$. To exercise all of the t-way combinations of inputs to a program, we need to cover all t-way combinations of variable values, and each combination of t values can have v^t configurations, where v is the number of values per variable. Thus, the total number of combinations instantiated with values that must be covered is

$$v^t \binom{n}{t} \tag{A.1}$$

Fortunately, each test covers $C(n, t)$ combination configurations. This fact is the source of combinatorial testing power. For example, with 34 binary variables, we would need $2^{34} = 1.7 \times 10^{10}$ tests to cover all possible configurations, but with only 33 tests we can cover all 3-way combinations of these 34 variables. This happens because each test covers $C(34, 3)$ combinations.

Example

If we have five binary variables, *a*, *b*, *c*, *d*, and *e*, then expression (A.1) says, we will need to cover $2^3 \times C(5, 3) = 8 \times 10 = 80$ configurations. For 3-way combinatorial testing, we will need to take all 3-variable combinations, of which there are 10:

abc, abd, abe, acd, ace, ade, bcd, bce, bde, cde

Each of these will need to be instantiated with all eight possible configurations of three binary variables:

000, 001, 010, 011, 100, 101, 110, 111

The test [0 1 0 0 1] covers the following C(5, 3) = 10 configurations:

abc abd abe acd ace ade bcd bce bde cde

010 000 011 001 001 001 100 101 101 001

A.1.2 Orthogonal Arrays

Some software testing problems can be solved with an orthogonal array, a structure that has been used for combinatorial testing in fields other than software for decades. An orthogonal array, $OA_\lambda(N;t,k,v)$ is an $N \times k$ array. In every $N \times t$ subarray, each *t*-tuple occurs exactly λ times. We refer to *t* as the *strength* of the coverage of interactions, *k* as the number of parameters or components (degree), and *v* as the number of possible values for each parameter or component (order). Other popular notations for orthogonal arrays include OA (N, v^k, t) and OA (N, k, v, t) (see Chapter 14).

Example

Suppose we have a system with three on–off switches, controlled by an embedded processor. Table A.1 tests all pairs of switch settings exactly once each. Thus, $t = 2$, $\lambda = 1$, $v = 2$. Note that there are $v^t = 2^2$ possible combinations of values for each pair: 00, 01, 10, 11. There are C(3, 2) = 3 ways to select switch pairs: (1,2), (1,3), and (2,3), and each test covers three pairs, so the four tests cover a total of 12 combinations which implies that each combination is covered exactly once. As one might suspect, it can be very challenging to fit all combinations to be covered into a set of tests exactly the same number of times.

TABLE A.1 Pairs of Switch Settings

Test	Sw 1	Sw 2	Sw 3
1	0	0	0
2	0	1	1
3	1	0	1
4	1	1	0

A.1.3 Covering Arrays

An alternative to an orthogonal array is a set called a *covering array*, which includes all t-way combinations of parameter values, for the desired strength t. A covering array, $CA_{\lambda}(N;t,k,v)$, is an $N \times k$ array. In every $N \times t$ subarray, each t-tuple occurs *at least* λ times. Note this distinction between covering arrays and orthogonal arrays discussed in the previous section. The covering array relaxes the restriction that each combination is covered exactly the same number of times. Thus, covering arrays may result in some test duplication, but they offer the advantage that they can be computed for much larger problems than is possible for orthogonal arrays. Software described elsewhere in this book can efficiently generate covering arrays up to strength $t = 6$, for a large number of variables.

The problems discussed in this publication deal only with the case when $\lambda = 1$ (i.e., that every t-tuple must be covered at least once). Other popular notations for such covering arrays include CA (N, v^k, t) and CA (N, k, v, t) (see Chapter 14). In software testing, each row of the covering array represents a test, with one column for each parameter that is varied in testing. Collectively, the rows of the array include every t-way combination of parameter values at least once. For example, Figure A.1 shows a

	Parameters									
Rows	**1**	**2**	**3**	**4**	**5**	**6**	**7**	**8**	**9**	**10**
1	0	0	0	0	0	0	0	0	0	0
2	1	1	1	1	1	1	1	1	1	1
3	1	1	1	0	1	0	0	0	0	1
4	1	0	1	1	0	1	0	1	0	0
5	1	0	0	0	1	1	1	0	0	0
6	0	1	1	0	0	1	0	0	1	0
7	0	0	1	0	1	0	1	1	1	0
8	1	1	0	1	0	0	1	0	1	0
9	0	0	0	1	1	1	0	0	1	1
10	0	0	1	1	0	0	1	0	0	1
11	0	1	0	1	1	0	0	1	0	0
12	1	0	0	0	0	0	0	1	1	1
13	0	1	0	0	0	1	1	1	0	1

FIGURE A.1 Three-way covering array for 10 parameters with 2 values each.

covering array that includes all 3-way combinations of binary values for 10 parameters. Each column gives the values for a particular parameter. It can be seen that any three columns in any order contain all eight possible combinations of the parameter values. Collectively, this set of tests will exercise all 3-way combinations of input values in only 13 tests, as compared with 1024 for exhaustive coverage.

A.1.4 Number of Tests Required

The challenge in computing covering arrays is to find the smallest possible array that covers all configurations of *t* variables. If every new test generated covered all previously uncovered combinations, then the number of tests needed would be

$$\frac{v^t C(n,t)}{C(n,t)} = v^t$$

Since this is not generally possible, the covering array will be significantly larger than v^t, but still a reasonable number for testing. It can be shown that the number of tests in a *t*-way covering array will be proportional to

$$v^t \log n \tag{A.2}$$

for *n* variables with *v* values each.

It is worth considering the components of this expression to gain a better understanding of what will be required to do combinatorial testing. First, note that the number of tests grows exponentially with the interaction strength *t*. The number of tests required for $t + 1$-way testing will be in the neighborhood of *v* times the number required for *t*-way testing. Table A.2 shows how v^t grows for values of *v* and *t*. Although the number of tests required for high-strength combinatorial testing can be very large, with advanced software and cluster processors may not be out of reach.

Despite the possibly discouraging numbers in Table A.2, there is some good news. Note that formula (A.2) grows only logarithmically with the

TABLE A.2 Growth of v^t

v↓ t→	2	3	4	5	6
2	4	8	16	32	64
4	16	64	256	1024	4096
6	36	216	1296	7776	46,656

number of variables, n. This is fortunate for software testing. Early applications of combinatorial methods typically involved small numbers of variables, such as a few different types of crops or fertilizers, but for software testing, we must deal with tens, or in some cases hundreds of variables.

A.2 REGULAR EXPRESSIONS

Regular expressions are formal descriptions of strings of symbols, which may represent text, events, characters, or other objects. They are developed within automata theory and formal languages, where it is shown that there are direct mappings between expressions and automata to process them, and are encountered in many areas in the field of computer science. In combinatorial testing they may be encountered in sequence covering or in processing test input or output. Implementations vary, but standard syntax is explained below.

A.2.1 Expression Operators

Basic elements of regular expressions include

```
| "or" alternation. Ex: ab|ac matches "ab" or "ac"
? 0 or 1 of the preceding element. Ex: ab?c matches
  "ac" or "abc"
* 0 or more of the preceding element. Ex: ab*
  matches "a", "ab", "abb", "abbb" etc.
+ 1 or more of the preceding element. Ex: ab+
  matches "ab", "abb", "abbb" etc.
() grouping. Ex: (abc|abcd) matches "abc" or "abcd"
. matches any single character. Ex: a.c matches
  "abc", "axc", "a@c" etc.
[] matches any single character within brackets. Ex:
  [abc] matches "a" or "b" or "c".
  A range may also be specified. Ex: [a-z] matches
  any single lower case character. (This option
  depends on the character set supported.)
[^] matches any single character that is not
  contained in the brackets.
  Ex: [^ab] matches any character except "a" or "b"
^ matches start position, that is, before the first
  character
$ matches end position, that is, after the last
  character
```

A.2.2 Combining Operators

The operators above can be combined with symbols to create arbitrarily complex expressions. Examples include

```
.*a.*b.*c.*  "a" followed by "b" followed by "c" with
             zero or more symbols prior to "a",
             following "c", or interspersed with the
             three symbols
a|b*         null or "a" or zero or more occurrences
             of "b"
a+           equivalent to aa*
```

Many regular expression utilities such as *egrep* support a broader range of operators and features. Readers should consult documentation for *grep*, *egrep*, or other regular expression processors for detailed coverage of the options available on particular tools.

Appendix B: Empirical Data on Software Failures

The Interaction Rule, introduced in Section 1.1, is based on empirical studies of failures in software. This section summarizes what is known about the distribution of *t*-way faults based on research by NIST (U.S. National Institute of Standards and Technology) and others [14,15,103–105,212]. Table B.1 summarizes what we know from studies of a variety of application domains (Table B.2), showing the percentage of failures that are triggered by the interaction of one to six variables. For example, 66% of the medical device failures were triggered by a single variable value, and 97% were triggered by either one or two variables interacting. Although certainly not conclusive, the available data suggest that the number of interactions involved in system failures is relatively low, with a maximum from 4 to 6 in the six studies cited below. Note, however, that the traffic collision avoidance system (TCAS) study used seeded errors, all others are "naturally occurring" (* = not reported). Seeded faults often have quite different characteristics than those introduced accidentally by programmers, which may account for the flatness of the curve for the TCAS detection rate. More recently, Zhang, Liu, and Zhang [224] found a maximum of 4-way interactions triggering failures in testing of commercial MP3 players, and the Document Object Model testing [221] reported in Chapter 2 also found a maximum of 4-way interactions needed to detect all errors.

We also have investigated a particular class of vulnerabilities, denial-of-serivce, using reports from the National Vulnerability Database (NVD), a publicly available repository of data on all publicly reported software

TABLE B.1 Percent Distribution of Variables Involved in Triggering Software Failures

Vars	Medical Devices	Browser	Server	NASA GSFC	Network Security	TCAS
1	66	29	42	68	17	*
2	97	76	70	93	62	53
3	99	95	89	98	87	74
4	100	97	96	100	98	89
5		99	96		100	100
6		100	100			

TABLE B.2 System Characteristics

System	System Type	Release Stage	Size (LOC)
Medical devices	Embedded	Fielded products	10^3–10^4 (varies)
Browser	Web browser	Development/beta release	Approx. 2×10^5
Server	HTTP server	Development/beta release	Approx. 10^5
NASA database	Distributed scientific database	Development, integration test	Approx. 10^5
Network security	Network protocols	Fielded products	10^3–10^5 (varies)

security vulnerabilities. NVD can be queried for fine-granularity reports on vulnerabilities. Data from 3045 denial-of-service vulnerabilities have the distribution shown in Table B.3. We present these data separately from that above because they cover only one particular kind of failure, rather than data on any failures occurring in a particular program as shown in Figure B.1.

Why do the failure detection curves look this way? That is, why does the error rate tail off so rapidly with more variables interacting? One

TABLE B.3 Cumulative Percentage of Denial-of-Service Vulnerabilities Triggered by *t*-Way Interactions

Vars	NVD Cumulative (%)
1	93
2	99
3	100
4	100
5	100
6	100

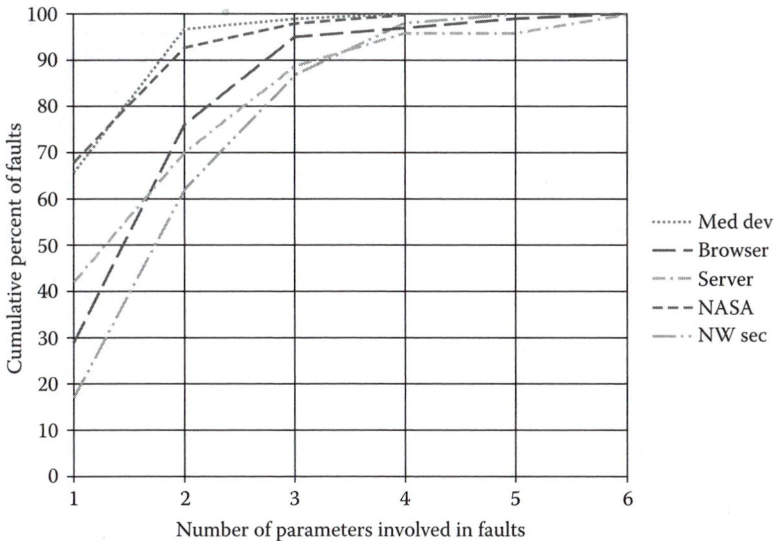

FIGURE B.1 (**See color insert.**) Cumulative percentage of failures triggered by t-way interactions.

possibility is that there are simply few complex interactions in branching points in software. If few branches involve 4-way, 5-way, or 6-way interactions among variables, then this degree of interaction could be rare for failures as well. Studies of code coverage with combinatorial testing tend to support this hypothesis. For example, one study [75] compared combinatorial and exhaustive testing on various measures of code coverage. Coverage was nearly identical and authors reported that only one segment of code was missed by 3-way testing, because it required a specific combination of four variables that would have been caught with a 4-way covering array. To see why combinatorial testing may produce high code coverage, consider the table above (Table B.4 and Figure B.2), which gives the number and percentage of branches in avionics code triggered by 1–19 variables. This distribution was developed by analyzing data in a report on the use of multiple condition decision coverage (MCDC) testing in avionics software [42], which included 20,256 logic expressions in five different airborne systems in two different airplane models. Table B.4 includes all 7685 expressions from *if* and *while* statements; expressions from assignment (:=) statements were excluded.

TABLE B.4 Number of Variables in Avionics Software Branches

Vars	Count	Percent	Cumulative (%)
1	5691	74.1	74.1
2	1509	19.6	93.7
3	344	4.5	98.2
4	91	1.2	99.3
5	23	0.3	99.6
6	8	0.1	99.8
7	6	0.1	99.8
8	8	0.1	99.9
9	3	0.0	100.0
15	1	0.0	100.0
19	1	0.0	100.0

As shown in Figure B.2, most branching statement expressions are simple, with over 70% containing only a single variable. Superimposing the curve from Figure B.2 on Figure B.1, we see (Figure B.3) that most failures are triggered by more complex interactions among variables. It is interesting that the NASA distributed database failures, from development-phase software bug reports, have a distribution similar to expressions in branching statements. This distribution may be because

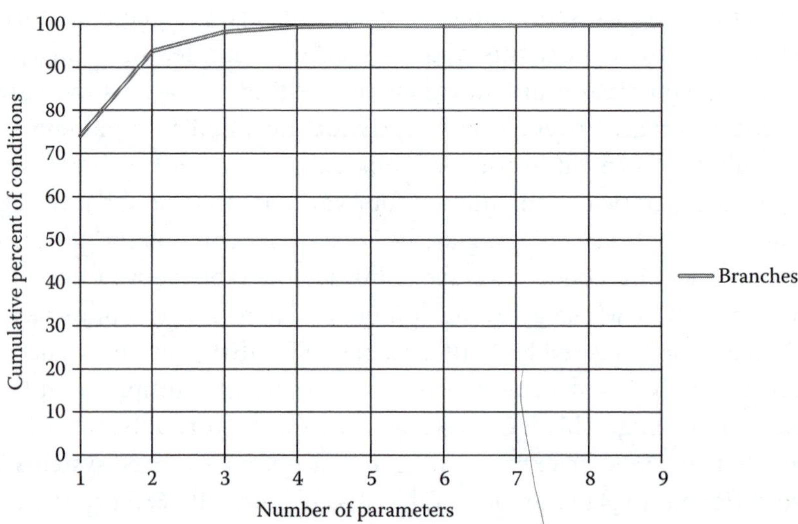

FIGURE B.2 **(See color insert.)** Cumulative percentage of branches containing *n* variables.

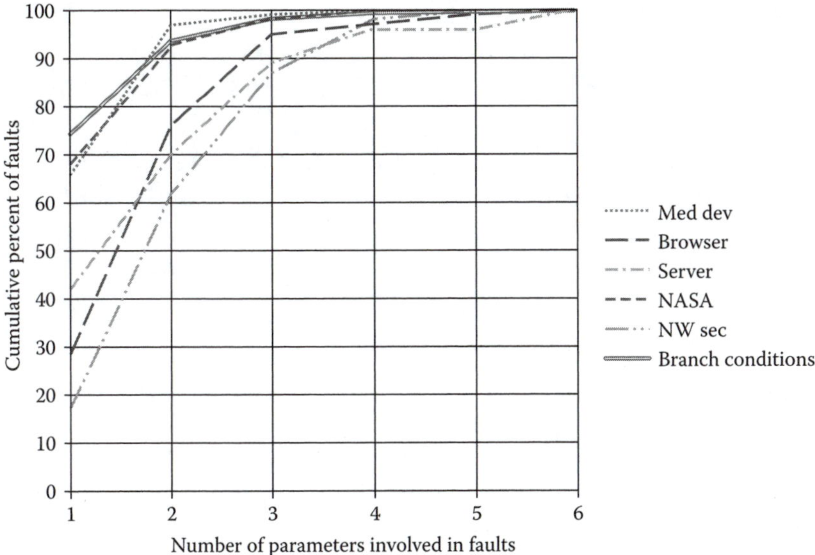

FIGURE B.3 **(See color insert.)** Branch distribution (green) superimposed in Figure B.1.

this was development phase rather than fielded software like all other types reported in Figure B.1. As failures are removed, the remaining failures may be harder to find because they require the interaction of more variables. Thus, testing and use may push the curve down and to the right.

Appendix C: Resources for Combinatorial Testing

A variety of software tools and resources are available to assist with combinatorial testing projects. Here, we summarize some that are freely available online. In addition to the tools given here, there are now many good commercial products that may have additional capabilities.

ACTS Web Site: http://csrc.nist.gov/acts. This is the NIST site that includes papers and the ACTS (automated combinatorial testing for software) and CCM (combinatorial coverage measurement) tool for download. The ACTS covering array generator is generally faster and produces smaller test arrays than others, based on comparisons done in 2009. The other tools available on this site, to the best of our knowledge, have functions that are not available elsewhere.

- ACTS covering array generator produces compact arrays that will cover 2-way through 6-way combinations. It also supports constraints that can make some values dependent on others, and mixed level covering arrays which offer different strength coverage for subsets of the parameters (e.g., 2-way coverage for one subset but 4-way for another subset of parameters). Output can be exported in a variety of formats, including human-readable, numeric, and spreadsheet. Either "don't care" or randomized output can be specified for tests that include combinations already fully covered by previous tests.
- CCM coverage measurement tool produces a comprehensive set of data on the combinatorial coverage of an existing set of tests, as

explained in Chapter 7. Output can be generated in spreadsheet format to allow easy processing and graphing.

- Sequence covering array generator produces sequence covering arrays as defined in Chapter 10. It includes an option for constraints in the form of prohibited sequences.

Best-Known Sizes of Covering Arrays: http://www.public.asu.edu/~ccolbou/src/tabby/catable.html. Professor Charles Colbourn, at Arizona State University, maintains data on the current best-known upper bounds on the smallest uniform covering array having k parameters (factors) each with v values (levels), with coverage at strength t, where $t = 2–6$. Sizes are reported for t, k, v, with k up to 20,000 for $t = 2$, and up to 10,000 for $t = 3–6$. Note that this database gives the best-known sizes only, and does not include the actual covering arrays.

Fault Evaluator: ECU (East Carolina University) Software Testing Research Group site. This tool can be used to evaluate the effectiveness of covering arrays or other test sets for detecting Boolean condition faults (see Section 4.3). Contact Sergiy Vilkomir through the STRG site: http://core.ecu.edu/STRG/

Combinatorial and Pairwise Testing Linked in Group: www.linkedin.com/groups/Combinatorial-Pairwise-Testing-4243185. This group is intended to facilitate discussion and sharing of information on combinatorial testing and its practical application to software assurance. Topics include questions/answers regarding test methods and tools, conference announcements, research, and experience reports.

Pairwise.org: www.pairwise.org. This site is a collection of links related to pairwise testing.

Testcover.com: www.testcover.com. Testcover.com offers a combinatorial test case generator for software engineers. Testcover.com is accessible via the cloud—a software as a service (SaaS) independent test design solution—and its Web Services Description Language (WSDL) interface enables ready integration with existing test tools.

CIT Lab: http://code.google.com/a/eclipselabs.org/p/citlab/. This open source Eclipse plugin provides a laboratory for combinatorial testing, featuring a standardized method and language for specifying combinatorial test problems, with syntax and semantics to allow exchange of models [73]. Also includes editor and development environment integrated with Eclipse.

Hexawise: http://www.hexawise.com. Online or tool-based combinatorial testing tools, with a large collection of training materials.

TOSCA: http://www.tricentis.com/en/tosca/automation/. A tool that offers a variety of features for combinatorial testing, including test automation, test case and test data management, and test case design [162].

Covering Array Tables: http://math.nist.gov/coveringarrays/. NIST maintains this collection of covering arrays computed by Michael Forbes' IPOG-F variant of the IPOG algorithm. Arrays are included for t, k, v, for $t = 2–6$, and $v = 2–6$ for t up to 5, and $v = 2–5$ for $t = 6$.

Sequence Covering Array Tables: http://csrc.nist.gov/groups/SNS/acts/sequence_cov_arrays.html

NIST maintains this collection of sequence covering arrays for 3-way and 4-way permutations of 5 through 100 events.

ACTS, CCM and sequence covering array generator are research tools not commercial products.

Appendix D: Test Tools

A VARIETY OF SOFTWARE TOOLS are available to assist with combinatorial testing projects. Here, we summarize the research tools available from the NIST website. The ACTS covering array generator is generally faster and produces smaller test arrays than others, based on comparisons we have done in 2009. The other tools listed below, to the best of our knowledge, have functions that are not available elsewhere.

All tools offered from the NIST website are research tools (not commercial products) and are offered for free download with no implied guarantees of performance and fitness for use. No claim against NIST can be made if downloaded a research tool from its website does not meet a user's expectations.

- ACTS covering array generator produces compact arrays that will cover 2-way through 6-way combinations. It also supports constraints that can make some values dependent on others, and mixed level covering arrays which offer different strength coverage for subsets of the parameters (e.g., 2-way coverage for one subset but 4-way for another subset of parameters). Output can be exported in a variety of formats, including human-readable, numeric, and spreadsheet. Either "don't care" or randomized output can be specified for tests that include combinations already fully covered by previous tests.

- Coverage measurement tool produces a comprehensive set of data on the combinatorial coverage of an existing set of tests, as explained in Chapter 7. Output can be generated in spreadsheet format to allow easy processing and graphing.

- Sequence covering array generator produces sequence covering arrays as defined in Chapter 10. It includes an option for constraints in the form of prohibited sequences.

To obtain any of these, see the ACTS web site at csrc.nist.gov/acts.

D.1 ACTS USER GUIDE

ACTS is a test generation tool for constructing t-way combinatorial test sets. Currently, it supports t-way test set generation with t ranging from 1 to 6. The ACTS tool is written in Java and provides API (application programming interface), command line and GUIs.

D.1.1 Core Features

D.1.1.1 t-Way Test Set Generation

ACTS supports t-way test set generation for $1 \leq t \leq 6$. Empirical studies show that t being up to 6 is sufficient for most practical applications. A special form of 1-way testing, called base-choice testing, is implemented in ACTS. Base-choice testing requires that every parameter value be covered at least once and in a test in which all the other values are base choices. Each parameter has one or more values designated as base choices. Informally, base choices are "more important" values, for example, default values, or values that are used most often in operation.

Several test generation algorithms are implemented in ACTS. These algorithms include IPOG, IPOG-D, IPOG-F, IPOG-F2, and PaintBall. In general, IPOG, IPOG-F, and IPOG-F2 work best for systems of moderate size (<20 parameters and 10 values per parameter on average), while IPOG-D and PaintBall are preferred for larger systems.

ACTS supports two test generation modes, namely, scratch and extend. The former allows a test set to be built from scratch, whereas the latter allows a test set to be built by extending an existing test set. In the extend mode, an existing test set can be a test set that is generated by ACTS, but is incomplete because of some newly added parameters and values, or because of a test set that is supplied by the user and imported into ACTS. Extending an existing test set can save earlier effort that has already been spent in the testing process.

D.1.1.2 Mixed Strength

This feature allows different parameter groups to be created and covered with different strengths. For example, consider a system consisting of 10 parameters, P1, P2,…, and P10. A default relation can be created that consists of all the parameters with strength 1 or 2. Then, additional relations can be created if some parameters are believed to have a higher degree of interaction, based on the user's domain knowledge. For instance, a relation could be created that consists of P2, P4, P5, P7 with strength 4 if the four parameters are closely related to each other, and their 4-way interactions could trigger certain software faults. ACTS allows arbitrary parameter relations to be created, where different relations may overlap or subsume each other. In the latter case, relations that are subsumed by other relations will be ignored by the test generation engine. Mixed strength is only supported in IPOG and IPOG-F algorithm.

D.1.1.3 Constraint Support

Some combinations are not valid from the domain semantics, and must be excluded from the resulting test set. For example, when we want to make sure a web application can run in different Internet browsers and on different Operating Systems, the configuration of IE on Mac OS is not a valid combination. A test that contains an invalid combination will be rejected by the system (if adequate input validation is performed) or may cause the system to fail. In either case, the test will not be executed properly, which may compromise test coverage, if some (valid) combinations are only covered by this test.

ACTS allows the user to specify constraints that combinations must satisfy to be valid. The specified constraints will be taken into account during test generation so that the resulting test set will cover, and only cover, combinations that satisfy these constraints. Currently, constraint support is only available for the IPOG and IPOG-F algorithm. Constraint support for other algorithms will be added in a future release.

D.1.1.4 Coverage Verification

This feature is used to verify whether a test set satisfies *t*-way coverage, that is, whether it covers all the *t*-way combinations. A test set to be verified can be a test set generated by ACTS or a test set supplied by the user (and then imported into ACTS).

D.1.2 Command Line Interface

There is a separate jar file for the command line version and for the GUI version. The command line version can be executed using the following command:

```
java <options> -jar acts_cmd.jar ActsConsoleManager <input_
  filename> <output_filename >
```

The various options are

```
-Dmode = scratch|extend
        scratch - generate tests from scratch (default)
        extend - extend from an existing test set

-Dalgo = ipog|ipog_d|bush|rec|paintball|ipof|ipof2|basechoice
        ipog - use algorithm IPO (default)
        ipog _ d - use algorithm IPO + Binary Construction (for
             large systems)
        bush - use Bush's method
        paintball - use the Paintball method
        ipof - use the IPOF method
        ipof2 - use the IPOF method
        basechoice - use Base Choice method

-DfastMode = on|off
        on - enable fast mode
        off - disable fast mode

-Ddoi = <int> :
        specify the degree of interactions to be covered

-Doutput = numeric|nist|csv|excel
        numeric - output test set in numeric format
        nist - output test set in NIST format (default)
        csv - output test set in CSV format
        excel - output test set in EXCEL format

-Dcheck = on|off:
        on - verify coverage after test generation
        off - do not verify coverage (default)

-Dprogress = on|off:
        on - display progress information (default)
        off - do not display progress information

-Dhunit = <int> :
        the number of tests extended during horizontal exten-
             sion per progress unit

-Dvunit = <int> :
        the number of pairs covered during vertical growth
             per progress unit

-Ddebug = on|off:
        on - display debug info
        off - do not display debug info (default)
```

```
-Drandstar=on|off:
        on—randomize don't care values
        off—do not randomize don't care values

-Dcombine=<all>:
    all - every possible combination of parameters
```

The above usage information can be displayed using the following command:

```
java -jar acts_cmd.jar
```

In the command line, <input_file> contains the configuration information of the system to be tested. The format of a configuration file is illustrated using the following example:

```
[System]
-- specify system name
Name: Test Configuration from Rick
[Parameter]
-- general syntax is parameter_name : value1, value2, ....
-- only compare with MINSEP and MAXALTDIFF general
Cur_Vertical_Sep (int) : 299, 300, 601
High_Confidence (boolean) : TRUE, FALSE
Two_of_Three_Reports_Valid (boolean) : TRUE, FALSE

-- Low and High, only compare with Other_Tracked_Alt
-- Own_Tracked_Alt (int) : 1, 2
Other_Tracked_Alt (int) : 1, 2

-- only compare with OLEV
Own_Tracked_Alt_Rate (enum) : 600, 601
Alt_Layer_Value (int) : 0, 1, 2, 3

-- compare with each other (also see NOZCROSS) and with
-- ALIM
Up_Separation (int) : 0, 399, 400, 499, 500, 639, 640,
   739, 740, 840
Down_Separation (int) : 0, 399, 400, 499, 500, 639, 640,
   739, 740, 840
Other_RAC (enum): NO_INTENT, DO_NOT_CLIMB, DO_NOT_DESCEND
Other_Capability (enum) : TCAS_TA, OTHER

Climb_Inhibit (boolean): TRUE, FALSE
```

```
[Relation]
-- this section is optional
-- general format Rx : (p1, p2, ..., pk,Strength)
R1 : (Cur_Vertical_Sep,Up_Separation,Down_Separation,3)
[Constraint]
-- this section is also optional
Cur_Vertical_Sep !=299 => Other_Capability != "OTHER"
Climb_Inhibit=true => Up_Separation>399
```

Currently, three parameter types are supported: enum, bool (boolean), and int (integer). Lines beginning with -- represent comments that exist only to improve the readability of the configuration file.

The default heap size for the Java Virtual Machine may not be adequate for large configurations. The user is recommended to change the default heap size, if necessary, using the following command:

```
java -Xms <initial heap size> -Xms<max heap size>
<options> -jar acts_cmd.jar ActsConsoleManager <input_file>
<output_file>
```

D.1.3 The GUI

There are two ways to launch the GUI front end. One way is to double-click the jar file for the GUI version, which is an executable jar file. The other way is to execute the jar file for the GUI version on the command prompt as follows:

```
java -jar acts_gui.jar
```

The following command can be used to change the default heap size for java virtual machine, if necessary:

```
java -Xms <initial heap size> -Xms<max heap size>
<options> -jar acts_gui.jar
```

Figures D.1 and D.2 show the general layout of the ACTS GUI. The System View component is a tree structure that shows the configurations

	CUR_...	HIGH_C...	TWO_OF...	OTH...	OWN_...	ALT_...	UP_S...	DOW...	OTHER_RAC	OTHER_CAP...	CLIMB...
1	300	FALSE	FALSE	2	601	1	0	0	DO_NOT_CLIMB	TCAS_TA	FALSE
2	299	TRUE	TRUE	1	600	3	0	399	NO_INTENT	OTHER	FALSE
3	601	FALSE	TRUE	2	600	2	0	400	DO_NOT_DESCEND	TCAS_TA	FALSE
4	299	FALSE	TRUE	1	601	0	0	499	DO_NOT_CLIMB	OTHER	FALSE
5	601	TRUE	FALSE	1	601	1	0	500	NO_INTENT	TCAS_TA	FALSE
6	601	TRUE	FALSE	2	600	3	0	639	DO_NOT_CLIMB	TCAS_TA	FALSE
7	299	TRUE	FALSE	1	601	2	0	640	DO_NOT_DESCEND	OTHER	FALSE
8	300	TRUE	TRUE	1	600	2	0	739	NO_INTENT	TCAS_TA	FALSE
9	300	FALSE	FALSE	2	601	3	0	740	DO_NOT_DESCEND	TCAS_TA	FALSE
10	299	FALSE	TRUE	2	600	1	0	840	DO_NOT_DESCEND	OTHER	FALSE
11	601	TRUE	FALSE	1	600	0	399	0	DO_NOT_DESCEND	TCAS_TA	FALSE
12	300	FALSE	FALSE	2	601	1	399	399	DO_NOT_CLIMB	TCAS_TA	FALSE
13	299	TRUE	TRUE	1	601	3	399	400	NO_INTENT	OTHER	FALSE
14	601	TRUE	FALSE	2	600	2	399	499	DO_NOT_DESCEND	TCAS_TA	FALSE
15	299	FALSE	TRUE	2	600	2	399	500	DO_NOT_CLIMB	OTHER	FALSE
16	300	FALSE	TRUE	1	601	2	399	639	DO_NOT_DESCEND	TCAS_TA	FALSE
17	300	FALSE	TRUE	2	600	3	399	640	DO_NOT_CLIMB	TCAS_TA	FALSE
18	601	FALSE	FALSE	2	601	1	399	739	DO_NOT_CLIMB	TCAS_TA	FALSE
19	299	TRUE	TRUE	2	600	0	399	740	NO_INTENT	OTHER	FALSE
20	300	TRUE	FALSE	1	601	0	399	840	NO_INTENT	TCAS_TA	FALSE
21	299	FALSE	TRUE	2	600	2	400	0	NO_INTENT	OTHER	TRUE
22	601	TRUE	FALSE	1	601	0	400	399	DO_NOT_DESCEND	TCAS_TA	TRUE
23	300	FALSE	FALSE	1	600	1	400	400	DO_NOT_CLIMB	TCAS_TA	TRUE
24	300	FALSE	TRUE	1	600	3	400	499	NO_INTENT	TCAS_TA	TRUE
25	300	TRUE	TRUE	2	601	0	400	500	DO_NOT_DESCEND	TCAS_TA	FALSE
26	299	TRUE	TRUE	2	600	0	400	639	NO_INTENT	TCAS_TA	TRUE
27	601	FALSE	TRUE	1	601	0	400	640	NO_INTENT	TCAS_TA	TRUE
28	299	TRUE	FALSE	2	600	0	400	739	DO_NOT_DESCEND	OTHER	TRUE
29	601	FALSE	FALSE	1	601	1	400	740	DO_NOT_CLIMB	TCAS_TA	TRUE
30	601	TRUE	FALSE	2	600	3	400	840	DO_NOT_CLIMB	TCAS_TA	TRUE
31	299	FALSE	TRUE	2	600	3	499	0	DO_NOT_DESCEND	TCAS_TA	FALSE
32	601	TRUE	FALSE	1	601	2	499	399	DO_NOT_CLIMB	TCAS_TA	TRUE

FIGURE D.1 The main window—test result tab.

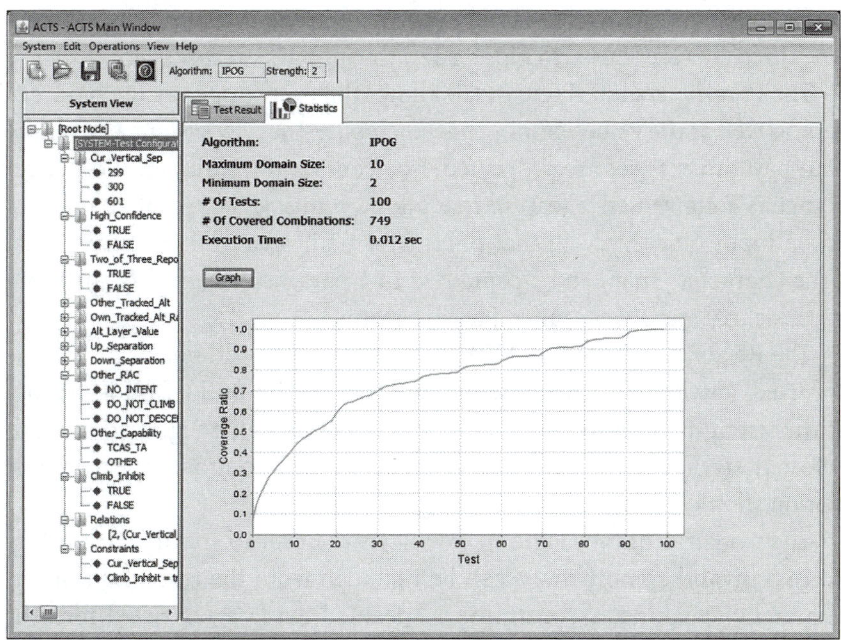

FIGURE D.2 The main window—statistics tab.

of the systems that are currently open in the GUI. In the tree structure, each system is shown as a three-level hierarchy. That is, each system (top level) consists of a set of parameters (second level), each of which has a set of values (leaf level). If a system has relations and constraints, they will be shown in the same level as the parameters.

To the right of the System View is a tabbed pane consisting of two tabs, namely, Test Result, which is shown in Figure D.1, and Statistics, which are shown in Figure D.2. The Test Result shows a test set of the currently selected system, where each row represents a test, and each column represents a parameter. Output parameters are also displayed as columns. The Statistics tab displays some statistical information about the test set. In particular, it includes a graph that plots the growth rate of the test coverage with respect to the tests in the test set displayed in the Test Result tab. Drawing the graph may involve expensive computations, and thus the graph is shown only on demand, that is, when the Graph button is clicked.

D.1.3.1 Create New System

To create a new system, select menu System -> New, or the first icon in the toolbar, to open the New System window. The New System window contains a tabbed pane of three tabs, namely, Parameters, Relations, and Constraints. The three tabs are shown in Figures D.3 through D.5, respectively.

The Parameters tab (Figure D.3) allows the user to specify the parameters, as well as the values of those parameters, in the new system. Currently, four parameter types are supported, Boolean, Enum, Number, and Range. Range is a convenience feature that allows multiple, consecutive integers to be input quickly. Note that parameter names cannot contain spaces. (The characters that can be contained in a parameter name are the same as those in a variable name in Java programs.)

The Relations tab (Figure D.4) allows the user to add one or more customized relations for the system. Note that in the Build Option dialog, if the strength is set to 1–6, the System relations will be ignored and the selected strength will apply. To use customized relations, the strength option should be set to "mixed strength."

The Constraints tab (Figure D.5) allows the user to specify constraints so that invalid combinations can be excluded from the resulting test set. Generally speaking, a constraint is specified using a restricted form of first-order logical formulas. In the following, we give a formal syntax of the expressions that can be used to specify a constraint:

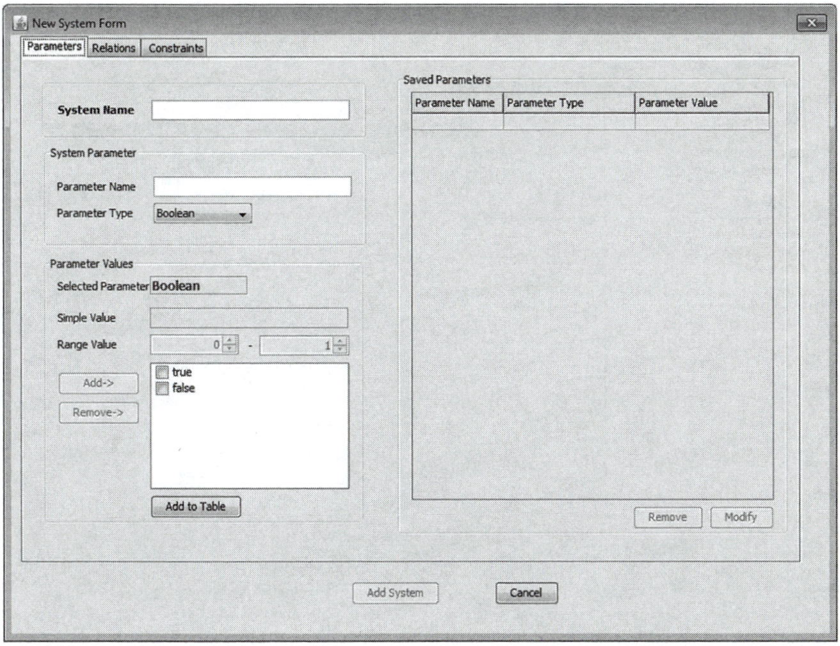

FIGURE D.3 New system window—parameters.

```
<Constraint > :: =<Simple _ Constraint > |< Constraint >
  <Boolean _ Op ><Constraint >
<Simple _ Constraint > :: =<Term><Relational _ Op ><Term>
<Term > :: =<Parameter > |< Parameter ><Arithmetic _ Op >
  <Parameter > | <Parameter ><Arithmetic_Op> <Value>
<Boolean_Op > : : = "&&"  |"||"  |"=>"
<Relational_Op> : : = "= "  |"! = "  | " > "  | " < "  | " >="  | " <="
<Arithmetic_Op> : : = "+ "  | "-"  | "*"  | "/"  | "%"
```

There are three types of operators: (1) Boolean operators (Boolean_Op), including &&, ||, =>; (2) Relational operators (Relational_Op), including =, !=, >, <, >=, <=; and (3) Arithmetic operators (Arithmetic_OP), including +, –, *, /, %. Note that arithmetic operators can appear in a term expression (<Term>) only if the parameters involved in the term expression are of type Number or Range. Also, four of the relational operators, namely, >, <, >=, <=, can appear in a simple constraint expression (Simple_Constraint) only if both of the terms involved in the simple constraint are evaluated to a parameter value of type Number or Range.

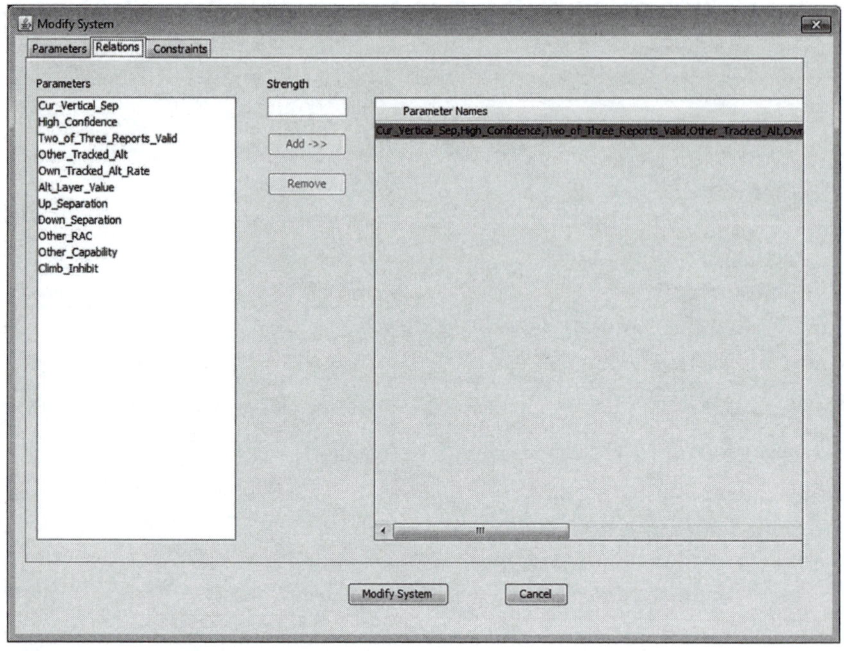

FIGURE D.4 New system window—relations.

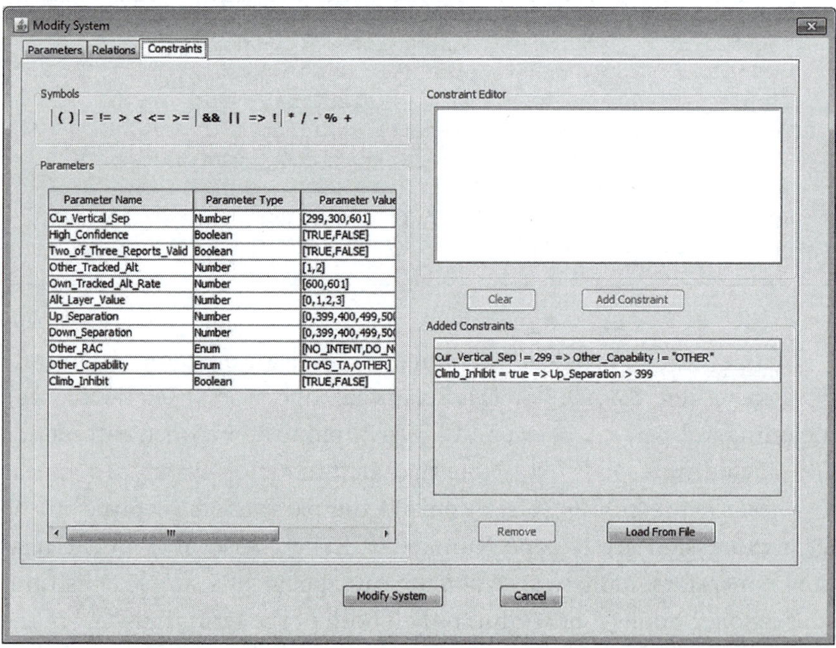

FIGURE D.5 New system window—constraints.

The following are examples of various constraints that can be specified:

- *Constraint 1*: (OS = "Windows") => (Browser = "IE" || Browser = "FireFox" || Browser = "Netscape"), where OS and Browser are two parameters of type Enum. This constraint specifies that if OS is Windows, then Browser has to be IE, FireFox, or Netscape.
- *Constraint 2*: (P1 > 100) || (P2 > 100), where P1 and P2 are two parameters of type Number or Range. This constraint specifies that P1 or P2 must be greater than 100.
- *Constraint 3*: (P1 > P2) => (P3 > P4), where P1, P2, P3, and P4 are parameters of type Number or Range. This constraint specifies that if P1 is greater than P2, then P3 must be greater than P4.
- *Constraint 4*: (P1 = true || P2 > =100) => (P3 = "ABC"), where P1 is a Boolean parameter, P2 is a parameter of type Number or Ranger, and P3 is of type Enum. This constraint specifies that if P1 is true and P2 is greater than or equal to 100, then P3 must be "ABC."

A constraint can be directly typed in the Constraint Editor. The user is provided with the system configuration and the operators that can be used. The left-hand side of the Constraint window displays the system configuration in a table format and the operators on the top of the system configuration table. Note that parameter values that are strings must be quoted in double quotes; otherwise, they will be considered as parameter names. The right bottom of this frame shows all the added constraints. By clicking the button "Load From File," one may load constraints from a plain txt file, in which each line contains a constraint expression. For example, a constraint file (in TXT format) contains two lines:

```
Cur_Vertical_Sep != 299 => Other_Capability != "OTHER"
Climb_Inhibit = true => Up_Separation > 399
```

An existing constraint can be removed by selecting the constraint in the Added Constraint table and then clicking on the Remove button. Currently, ACTS does not allow an existing constraint to be directly edited. To edit an existing constraint, the user needs to remove the constraint first and then add the desired constraint as a new constraint.

D.1.3.2 Build Test Set

To build a test set for a system that is currently open, select the system in the System View, and then select menu Operations -> Build. The latter selection brings up the Options window, as shown in Figure D.6, which allows the following options to be specified for the build operation:

- *Algorithm:* This option decides which algorithm to be used for test generation. As mentioned in Section D.1.1, IPOG, IPOG-F, and IPOG-F2 work best for systems of moderate size, while IPOG-D and PaintBall are preferred for larger systems. Note that relations and constraints are only supported for IPOG and IPOG-F. By default, the IPOG algorithm is selected.
- *Max Tries:* This option is used by algorithm PaintBall, and it specifies the number of candidates to be generated randomly at each step.
- *Randomize Don't Care Values:* If this option is checked, then all the Don't Care values (*) in the resulting test set will be replaced with a random value. The coverage will not be affected if we replace a Don't Care value with a real value.
- *Ignore Constraints:* If this option is checked, all constraints will be ignored. For some generation algorithms which do not support constraints, this option will be automatically checked.

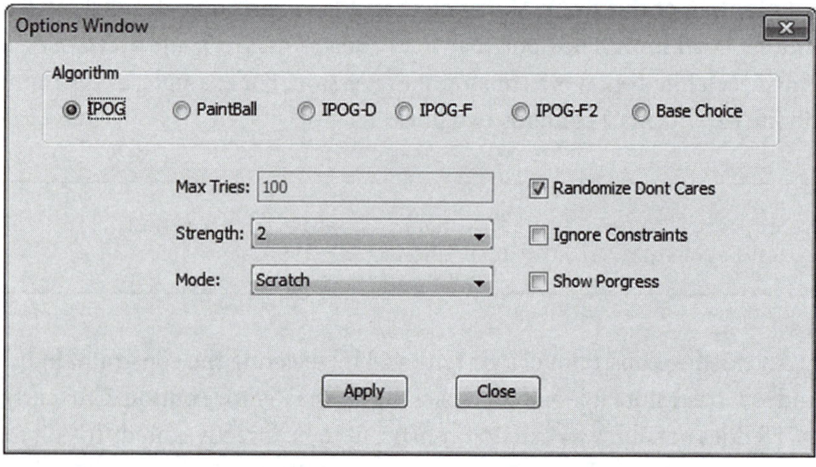

FIGURE D.6 Build options window.

- *Strength:* This option specifies the strength of the test set. Currently, ACTS supports a strength value ranging from 1 to 6. Note that the user may create customized relations. To use customized relations, select "mixed strength" from this drop-down list. The strength for the Base Choice algorithm always be 1.
- *Mode:* This option can be Scratch or Extend. The former specifies that a test set should be built from scratch; the latter specifies that a test set should be built by extending an existing test set (shown in the Test Result tab). Recall that the current test set in the system may not be complete as the system configuration may have changed after the last build or the test set may be imported from outside.
- *Progress:* If this option is turned on, progress information will be displayed in the console. Note that in order to obtain the console, the GUI must be started from a command prompt instead of by double-clicking the executable jar file.

After the build operation is completed, the resulting test set will be displayed in the Test Result tab of the Main window.

D.1.3.3 Modify System

To modify an existing system, select the system in the tree view, and then select menu Edit -> Modify. As shown in Figure D.7, the Modify System window is the same as the New System window except that the name of the system cannot be changed. A parameter cannot be removed if it is involved in a relation other than the default relation or constraint. In this case, the parameter must be removed from the relation or constraint first.

A parameter can be added in the same way as during the New System operation. A parameter can be removed by selecting the parameter in the Saved Parameters table on the right-hand side, and then clicking on the Remove button under the table. The values of a parameter can be modified by selecting the parameter on the Saved Parameters table on the right-hand side, and by clicking on the Modify button under the table. This brings up the Parameter Modification window, which is shown in Figure D.8.

A system can also be modified through the tree view. For example, a parameter, or value, or relation, or constraint can be removed by first selecting the parameter, or value, or relation, or constraint, and then selecting menu Edit -> Delete.

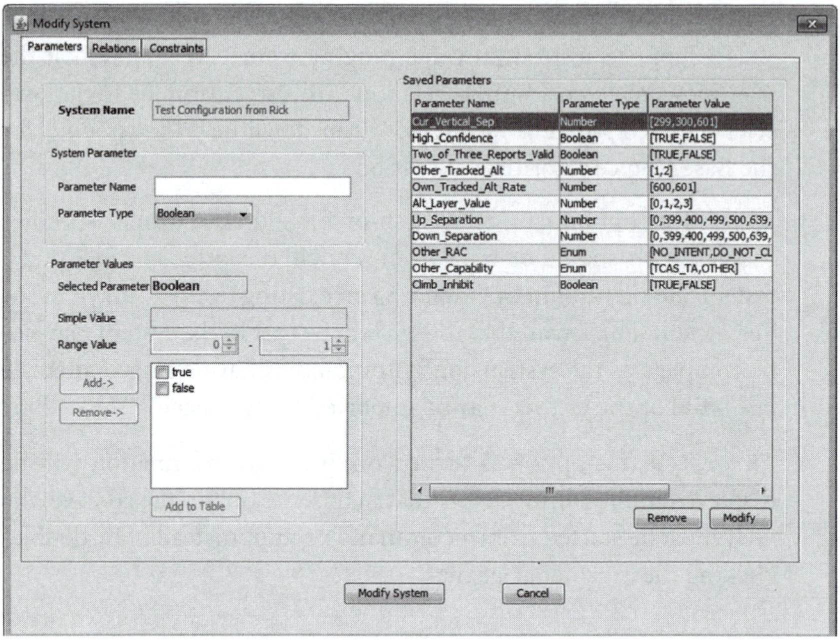

FIGURE D.7 Modify system window.

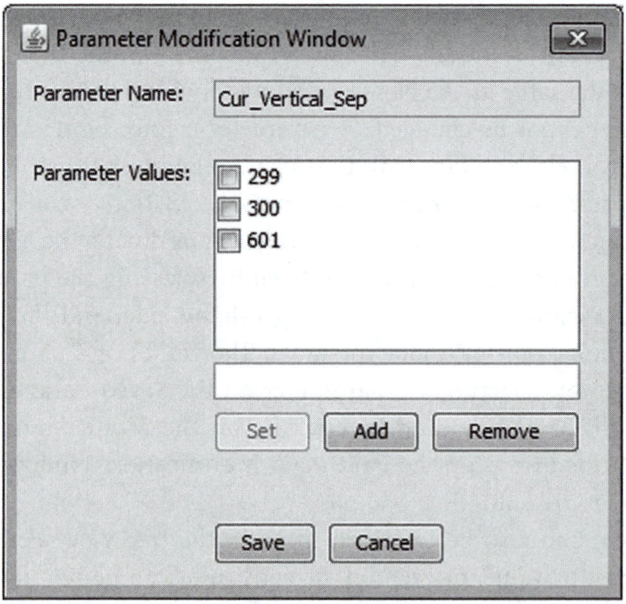

FIGURE D.8 Parameter modification window.

D.1.3.4 Save/Save as/Open System

To save an existing system, select the system in the tree view, and then select menu System -> Save or Save As. When a newly created system is saved for the first time, or when Save As is selected, a standard file dialog will be brought up, where the user can specify the name of the file to be saved. The system will display a confirmation window if the file to be saved already exists. Note that if you open a system from a TXT file and click "Save," the content will be overwritten in the XML format. So always use "Save As" for TXT files.

D.1.3.5 Import/Export Test Set

To import a test set of a system, the user must first create the system, in terms of adding its parameters and values into ACTS, as described in Section D.1.3.1. Then, select menu Operations -> Import, and select the format of the file containing the test set. Currently, two file formats are supported: CSV-R, which stands for Comma Separated Values with Row headers, and CSV-RC, which stands for Comma Separated Values with Row and Column headers. (CSV-RC is mainly used to facilitate integration with Excel.) The following are two example files, one for each format:

```
CSV-R format:

P1,P2,P3,P4,P5
0,2,2,3,6
3,2,4,2,2
2,1,2,1,3
3,2,5,0,5
```

```
CSV-RC format:

P1,P2,P3,P4,P5
Test1,0,2,2,3,6
Test2,3,2,4,2,2
Test3,2,1,2,1,3
Test4,3,2,5,0,5
```

The parameter values in each row must be separated by ",". There can be arbitrary space between two values. After the file format is selected, a standard file selection window appears through which the user can browse through the system and select the file containing the test set to be imported.

To export a test set that exists in the GUI, first select the corresponding system so that the test set is displayed in the Test Result tab of the Main window, and then select Operations -> Export. Currently, three formats are supported, namely, NIST Format, Excel Format, and CSV Format.

D.1.3.6 Verify t-Way Coverage

To verify *t*-way coverage of a test set, the user can select menu Operations -> Options, and specify a desired strength in the Options window. Then, select menu Operations -> Verify. If the test set achieves the coverage for the specified strength, a message window will be displayed; otherwise, another message window will show how many tuples are missing. Note that this operation is typically used to verify the coverage of a test set that is imported from outside of ACTS.

References

1. Ammann, P. and P. E. Black. Abstracting formal specifications to generate software tests via model checking. In *Proceedings of the 18th Digital Avionics Systems Conference*, Vol. 2, pp. 10-A. St. Louis, Missouri, IEEE, 1999.
2. Ammann, P. E. and J. C. Knight. Data diversity: An approach to software fault tolerance. *IEEE Transactions on Computers*, 37(4), 1988: 418–425.
3. Ammann, P. E. and A. J. Offutt. Using formal methods to derive test frames in category-partition testing. In *Proceedings of the Ninth Annual Conference on Computer Assurance (COMPASS'94)*, IEEE Computer Society Press, Gaithersburg, Maryland, pp. 69–80, 1994.
4. Ammann, P. and J. Offutt. *Introduction to Software Testing*. Cambridge University Press, New York, 2008.
5. Apilli, B. S., L. Richardson, and C. Alexander. Fault-based combinatorial testing of web services. In *Proceedings of the 24th ACM SIGPLAN Conference Companion on Object Oriented Programming Systems Languages and Applications*, Orlando, October 25–29, 2009.
6. Arcuri, A. and L. Briand. Adaptive random testing: An illusion of effectiveness? In *Proceedings of the 2011 International Symposium on Software Testing and Analysis*, pp. 265–275. New York, NY, ACM, 2011.
7. Arcuri, A. and L. Briand. Formal analysis of the probability of interaction fault detection using random testing. *IEEE Transactions on Software Engineering*, IEEE Computer Society, http://doi.ieeecomputersociety.org/10.1109/TSE.2011.85, August 18, 2011.
8. Bach, J. and P. J. Schroeder. Pairwise testing: A best practice that isn't. In *Proceedings of the 22nd Pacific Northwest Software Quality Conference*, pp. 180–196. Portland, Oregon, 2004.
9. Ballance, W. A., S. Vilkomir, and W. Jenkins. Effectiveness of pair-wise testing for software with Boolean inputs. In *Software Testing, Verification and Validation (ICST), 2012 IEEE Fifth International Conference*, pp. 580–586. Montreal, Canada, IEEE, 2012.
10. Ballance, W., W. Jenkins, and S. Vilkomir. Probabilistic assessment of effectiveness of software testing for safety-critical systems. In *Proceedings of the 10th International Probabilistic Safety Assessment and Management Conference, (PSAM'10)*, Seattle, Washington, 2010.
11. Banbara, M., N. Tamura, and K. Inoue. Generating event-sequence test cases by answer set programming with the incidence matrix. In *Technical*

Communications of the 28th International Conference on Logic Programming (ICLP'12), Vol. 17, pp. 86–97. Istanbul, Turkey, Schloss Dagstuhl—Leibniz-Zentrum fuer Informatik, 2012.

12. Baresi, L. and M. Young. Test oracles. *Technical Report CISTR-012*, 2001.
13. Beizer, B. *Software Testing Techniques*, 2nd edition. Van Nostrand Reinhold, New York, 1990.
14. Bell, K. Z. and M. A. Vouk. On effectiveness of pairwise methodology for testing network-centric software. In *Information and Communications Technology, Enabling Technologies for the New Knowledge Society: ITI 3rd International Conference*, pp. 221–235. Karachi, Pakistan, IEEE, 2005.
15. Bell, K. Z. Optimizing effectiveness and efficiency of software testing: A hybrid approach. PhD dissertation, North Carolina State University, 2006.
16. Binder, R. V. State machines. In *Testing Object-Oriented Systems, Models, Patterns, and Tools*, Chapter 7, pp. 175–268. Addison-Wesley, Reading, Massachusetts, 1999.
17. Bishop, P. G. The variation of software survival times for different operational input profiles. In *Proceedings of the 23rd International Symposium on Fault-Tolerant Computing (FTCS-23)*. IEEE Computer Society Press. pp. 98–107. Toulouse, France, 1993.
18. Black, P. E., V. Okun, and Y. Yesha. Testing with model checkers: Insuring fault visibility. *WSEAS Transaction Systems*, 2(1), 2003: 77–82.
19. Black, P. E., V. Okun, and Y. Yesha. Mutation operators for specifications. In *Automated Software Engineering, Proceedings ASE 2000, The Fifteenth IEEE International Conference on*, pp. 81–88. Grenoble, France, IEEE, 2000.
20. Boehm, B. W. *Software Engineering Economics*. Prentice-Hall, Upper Saddle River, New Jersey, 1981.
21. Borazjany, M., L. Yu, Y. Lei, R. Kacker, and D. R. Kuhn. Combinatorial testing of ACTS, a case study. *Workshop on Combinatorial Testing* (CT), *International Conference on Software Testing (ICST)*, Montreal, Canada, April 17, 2012.
22. Box, G. E. P., W. G. Hunter, and J. S. Hunter. An introduction to design, data analysis, and model building. *Statistics for Experimenters*, pp. 374–434. Wiley, Hoboken, New Jersey, 1978.
23. Brownlie, R., J. Prowse, and M. S. Phadke. Robust testing of AT&T PMX/StarMAIL using oats. *AT&T Technical Journal*, 71(3), 1992: 41–47.
24. Bryce, R. C. and C. J. Colbourn. The density algorithm for pairwise interaction testing. *Software Testing, Verification and Reliability*, 17(3), 2007: 159–182.
25. Bryce, R. C., C. J. Colbourn, and M. B. Cohen. A framework of greedy methods for constructing interaction tests. In *The 27th International Conference on Software Engineering (ICSE)*, St. Louis, Missouri, pp. 146–155, May 2005.
26. Bryce, R. C. and C. J. Colbourn. Prioritized interaction testing for pair-wise coverage with seeding and constraints. *Information and Software Technology*, 48(10), 2006: 960–970.
27. Bryce, R. C., A. Rajan, and M. P. E. Heimdahl. Interaction testing in model-based development: Effect on model-coverage. In *Software Engineering Conference, APSEC, 13th Asia Pacific*, pp. 259–268. Bangalore, India, IEEE, 2006.

28. Bryce, R. C., S. Sampath, and A. M. Memon. Developing a single model and test prioritization strategies for event-driven software. *IEEE Transactions on Software Engineering*, 37(1), 2011: 48–64.
29. Bryce, R. C. and A. M. Memon. Test suite prioritization by interaction coverage. In *Proceedings of the Workshop on Domain-Specific Approaches to Software Test Automation (DoSTA), 6th Joint Meeting of the European Software Engineering Conference and the ACM SIGSOFT Symposium on the Foundations of Software Engineering*, pp. 1–7. Dubrovnik, Croatia, September 2007.
30. Bryce, R. C., S. Sampath, J. Pedersen, and S. Manchester. Test suite prioritization by cost-based combinatorial interaction coverage. *International Journal on Systems Assurance Engineering and Management (Springer)*, 2(2), 2011: 126–134.
31. Bryce, R. C., Y. Lei, D. R. Kuhn, and R. Kacker. Combinatorial testing. In *Handbook of Research on Software Engineering and Productivity Technologies: Implications of Globalization*, Chapter 14, Ramachandran, ed., IGI Global, Hershey, Pennsylvania, 2009.
32. Burners-Lee, T. *Uniform Resource Locators (URL)*. Retrieved June 2012, from IETF: http://www.ietf.org/rfc/rfc1738.txt, 1994.
33. Burr, K. and W. Young. Combinatorial test techniques: Table-based automation, test generation, and test coverage, *International Conference on Software Testing, Analysis, and Review (STAR)*, San Diego, California, October 1998.
34. Burroughs, K., A. Jain, and R. L. Erickson. Improved quality of protocol testing through techniques of experimental design. In *Proceedings of the IEEE International Conference on Communications (Supercomm/ICC'94)*, pp. 745–752. New Orleans, Louisiana, IEEE, May 1994.
35. Bush, K. A. Orthogonal arrays of index unity. *Annals of Mathematical Statistics*, 23, 1952: 426–434.
36. Butler, R. W. and G. B. Finelli. The infeasibility of quantifying the reliability of life-critical real-time software. *IEEE Transactions on Software Engineering*, 19(1), 1993: 3–12.
37. Chen, T., H. Leung, and I. Mak. Adaptive random testing. In *Advances in Computer Science-ASIAN 2004. Higher-Level Decision Making*, pp. 3156–3157. Springer, New York, 2005.
38. Chen, T. Y. Adaptive random testing. In *Eighth International Conference on Quality Software*, pp. 443. Oxford, UK, IEEE, 2008.
39. Chen, T. Y., F.-C. Kuo, R. G. Merkel, and T. H. Tse. Adaptive random testing: The art of test case diversity. *Journal of Systems and Software*, 83(1), 2010: 60–66.
40. Chen, B., J. Yan, and J. Zhang. Combinatorial testing with shielding parameters. In *Software Engineering Conference (APSEC), 2010 17th Asia Pacific*, pp. 280–289. Sydney, Australia, IEEE, 2010.
41. Chen, B. and J. Zhang. Tuple density: A new metric for combinatorial test suites. In *Proceedings of the 33rd International Conference on Software Engineering*, pp. 21–28. Honolulu, Hawaii, 2011.

42. Chilenski, J. J. An investigation of three forms of the modified condition decision coverage (MCDC) criterion. Boeing Commercial Airplane Co, Seattle, Washington, 2001.
43. Cimatti, A., E. Clarke, F. Giunchiglia, and M. Roveri. NuSMV: A new symbolic model verifier. In *Computer Aided Verification*, pp. 682–682. Springer, Berlin, 1999.
44. Ciupa, I., A. Leitner, M. Oriol, and B. Meyer. ARTOO. In *Software Engineering, ICSE'08. 30th International Conference on ACM/IEEE,* pp. 71–80. Leipzig, Germany, IEEE, 2008.
45. Clarke, E., K. McMillan, S. Campos, and V. Hartonas-Garmhausen. Symbolic model checking. In *Computer Aided Verification*, pp. 419–422. Springer, Berlin, 1996.
46. Clarke, L. A., J. Hassell, and D. J. Richardson. A close look at domain testing. *IEEE Transactions on Software Engineering,* 4, 1982: 380–390.
47. Cochran, W. G. and M. G. Cox. *Experimental Designs*. Wiley, New York, 1950.
48. Codd, E. F. Further normalization of the data base relational model. *Data Base Systems,* pp. 33–64. Prentice Hall, Upper Saddle River, New Jersey, 1972.
49. Cohen, M. B., J. Snyder, and G. Rothermel. Testing across configurations: Implications for combinatorial testing. *ACM SIGSOFT Software Engineering Notes,* 31(6), 2006: 1–9.
50. Cohen, D. M., S. R. Dalal, J. Parelius, and G. C. Patton. The combinatorial design approach to automatic test generation. *Software IEEE,* 13(5), 1996: 83–88.
51. Cohen, D. M., S. R. Dalal, A. Kajla, and G. C. Patton. The automatic efficient test generator (AETG) system. In *Software Reliability Engineering, Proceedings of the 5th International Symposium*, pp. 303–309. Monterey, California, IEEE, 1994.
52. Colbourn, C. J. and J. H. Dinitz, eds. *Handbook of Combinatorial Designs*, Vol. 42. Chapman & Hall/CRC, Boca Raton, Florida, 2006.
53. Colbourn, C. J. Combinatorial aspects of covering arrays. *Le Matematiche (Catania),* 58, 2004: 121–167.
54. Colbourn, C. J., M. B. Cohen, and R. C. Turban. A deterministic density algorithm for pairwise interaction coverage. In *Software Engineering*. ACTA Press, Calgary, Canada, 2004.
55. Colbourn, C. J. Best known covering array sizes. http://www.public.asu.edu/~ccolbou/src/tabby/catable.html.
56. Colbourn, C. J. and D. W. McClary. Locating and detecting arrays for interaction faults. *Journal of Combinatorial Optimization,* 15(1), 2008: 17–48.
57. Copeland, L. *A Practitioner's Guide to Software Test Design*. Artech House Publishers, London, UK, 2004.
58. Cohen, M. B., M. B. Dwyer, and J. Shi. Interaction testing of highly-configurable systems in the presence of constraints. In *Proceedings of the 2007 International Symposium on Software Testing and Analysis*, pp. 129–139. London, UK, ACM, 2007.
59. Cohen, M. B., M. B. Dwyer, and J. Shi. Coverage and adequacy in software product line testing. In *Proceedings of the ISSTA 2006 Workshop on Role of Software Architecture for Testing and Analysis*, pp. 53–63. Portland, Maine, ACM, 2006.

60. Cohen, D. M., S. R. Dalal, J. Parelius, and G. C. Patton. The combinatorial design approach to automatic test generation. *IEEE Software*, 13(5), 1996: 83–87.
61. Cohen, D. M., S. R. Dalal, M. L. Fredman, and G. C. Patton. The AETG system: An approach to testing based on combinatorial design. *IEEE Transactions on Software Engineering*, 23(7), 1997: 437–444.
62. Czerwonka, J. (webpage) http://www.pairwise.org/. Last updated in July 2012 and accessed on April 23, 2013.
63. Czerwonka, J. Pairwise testing in the real world: Practical extensions to test-case scenarios. In *Proceedings of the Twenty-fourth Annual Pacific Northwest Software Quality Conference*, pp. 419–430. Portland, Oregon, PNSQC, 2006.
64. Dalal, S. R. and C. L. Mallows. Factor-covering designs for testing software. *Technometrics,* 40(3), 1998: 234–243.
65. Dunietz, I. S., W. K. Ehrlich, B. D. Szablak, C. L. Mallows, and A. Iannino. Applying design of experiments to software testing: Experience report. In *Proceedings of the 19th International Conference on Software Engineering*, pp. 205–215. Boston, Massachusetts, ACM, 1997.
66. du Bousquet, L., Y. Ledru, O. Maury, C. Oriat, and J-L. Lanet. A case study in JML-based software validation. In *Automated Software Engineering: Proceedings of the 19th IEEE International Conference on Automated Software Engineering*, Vol. 20(24), pp. 294–297. Linz, Austria, 2004.
67. Erdem, E., K. Inoue, J. Oetsch, J. Pührer, H. Tompits, and C. Yilmaz. Answer-set programming as a new approach to event-sequence testing. In *VALID 2011, the Third International Conference on Advances in System Testing and Validation Lifecycle*, pp. 25–34, Barcelona, Spain, 2011.
68. Fisher, R. A. *Statistical Methods for Research Workers.* Vol. 14. Oliver and Boyd, Great Britain, Edinburgh, 1970.
69. Fisher, R. A. *The Design of Experiments.* Oliver and Boyd, Great Britain, Edinburgh, 1935.
70. Finelli, G. B. NASA software failure characterization experiments. *Reliability Engineering & System Safety*, 32(1), 1991: 155–169.
71. Forbes, M., J. Lawrence, Y. Lei, R. N. Kacker, and D. R. Kuhn. Refining the in-parameter-order strategy for constructing covering arrays. *Journal of Research of NIST*, 113, 2008: 287–297.
72. Forbes, M. (webpage) http://math.nist.gov/coveringarrays/. Last update April 17, 2013 and accessed on April 23, 2013.
73. Gargantini, A. and P. Vavassori. CITLAB: A laboratory for combinatorial interaction testing. In *Software Testing, Verification and Validation (ICST), IEEE Fifth International Conference*, pp. 559–568. IEEE, Montreal, Canada, 2012.
74. Ghandehari, L., S. Gholamhossein, Y. Lei, T. Xie, R. Kuhn, and R. Kacker. Identifying failure-inducing combinations in a combinatorial test set. In *Software Testing, Verification and Validation (ICST), IEEE Fifth International Conference*, pp. 370–379. Montreal, Canada, IEEE, 2012.
75. Giannakopoulou, D., D. H. Bushnell, J. Schumann, H. Erzberger, and K. Heere. Formal testing for separation assurance. *Annals of Mathematics and Artificial Intelligence,* 63(1), 2011: 5–30.

76. Grindal, M., J. Offutt, and S. F. Andler. Combination testing strategies: A survey. *Software Testing, Verification and Reliability,* 15(3), 2005: 167–199.
77. Grindal, M., J. Offutt, and J. Mellin. Managing conflicts when using combination strategies to test software. In *18th Australian Software Engineering Conference, ASWEC,* pp. 255–264. Melbourne, Australia, IEEE, 2007.
78. Grochtmann, M. Test case design using classification trees. In *Proceedings of STAR,* Vol. 94, pp. 93–117. Washington, D.C., 1994.
79. Grochtmann, M. and K. Grimm. Classification trees for partition testing. *Software Testing, Verification and Reliability,* 3(2), 2006: 63–82.
80. Guo, Y. and S. Sampath. Web application fault classification—An exploratory study. In *Proceedings of the Second ACM-IEEE International Symposium on Empirical Software Engineering and Measurement,* pp. 303–305. Banff, Canada, ACM, 2008.
81. Cunningham, A. M., J. Hagar, and R. J. Holman. A system analysis study comparing reverse engineered combinatorial testing to expert judgment. In *IEEE Fifth International Conference on Software Testing, Verification and Validation (ICST),* pp. 630–635. Montreal, Canada, IEEE, 2012.
82. Harel, D. Statecharts: A visual formalism for complex systems. *Science of Computer Programming,* 8(3), 1987: 231–274.
83. Hartman, A. Software and hardware testing using combinatorial covering suites. *Graph Theory, Combinatorics and Algorithms,* pp. 237–266, Springer, New York, 2005.
84. Hartman, A. and L. Raskin. Problems and algorithms for covering arrays. *Discrete Mathematics,* 284(1), 2004: 149–156.
85. Hedayat, A. S., N. James, A. Sloane, and J. Stufken. *Orthogonal Arrays: Theory and Applications.* Springer, New York, 1999.
86. Hoare, C. A. R. Assertions: A personal perspective. *Annals of the History of Computing, IEEE* 25(2), 2003: 14–25.
87. Jaaksi, A. Developing mobile browsers in a product line. *Software, IEEE,* 19(4), 2002: 73–80.
88. Kacker, R. N. Off-line quality control, parameter design, and the Taguchi method. *Journal of Quality Technology,* 17, 1987: 176–188.
89. Kauppinen, R. and J. Taina. Rita environment for testing framework-based software product lines. In *Proceedings of the Eighth Symposium on Programming Languages and Software Tools (SPLST'2003),* pp. 58–69. Kuopio, Finland, 2003.
90. Kim, C., D. S. Batory, and S. Khurshid. Reducing combinatorics in testing product lines. In *Proceedings of the Tenth International Conference on Aspect-Oriented Software Development,* pp. 57–68. Porto de Galinhas, Brazil, ACM, 2011.
91. Kamsties, E. and C. Lott. An empirical evaluation of three defect detection techniques, Technical Report ISERN 95-02, Department of Computer Science, University of Kaiserslauten.
92. Kamsties, E. and C. Lott. An empirical evaluation of three defect-detection techniques. *Software Engineering—ESEC'95,* 1995: 362–383.
93. Kempthorne, O. *Design and Analysis of Experiments.* Wiley, New York, 1952.

94. Kleitman, D. J. and J. Spencer. Families of k-independent sets. *Discrete Mathematics*, 6, 1973: 255–262.
95. Kramer, R. iContract—The Java design by contract tool. In *Proceedings of TOOLS26: Technology of Object-Oriented Languages and Systems*, pp. 295–307. Santa Barbara, California, IEEE, 1998.
96. Hu, V. C., D. R. Kuhn, and T. Xie. Property verification for generic access control models. In *IEEE/IFIP International Conference on Embedded and Ubiquitous Computing, 2008. EUC'08*, Vol. 2, pp. 243–250. Shanghai, China, IEEE, 2008.
97. Institute of Electrical and Electronics Engineers, *IEEE Standard Glossary of Software Engineering Terminology*, ANSI/IEEE Std. pp. 729–1983.
98. Krueger, C. W. New methods in software product line practice. *Communications of the ACM*, 49(12), 2006: 37–40.
99. Kruse, P. M. and K. Lakhotia. Multi objective algorithms for automated generation of combinatorial test cases with the classification tree method. In *Symposium on Search Based Software Engineering (SSBSE 2011)*, Szeged, Hungary, 2011.
100. Kuhn, D. R. Fault classes and error detection capability of specification-based testing. *ACM Transactions on Software Engineering and Methodology (TOSEM)*, 8(4), 1999: 411–424.
101. Kuhn, R., R. Kacker, Y. Lei, and J. Hunter. Combinatorial software testing. *Computer*, 42(8), 2009: 94–96.
102. Kuhn, D. R., R. Kacker, and Y. Lei. Automated combinatorial test methods—Beyond pairwise testing. *Crosstalk, Journal of Defense Software Engineering*, 21(6), 2008: 22–26.
103. Kuhn, D. R. and V. Okun. Pseudo-exhaustive testing for software. In *Software Engineering Workshop, SEW'06. 30th Annual IEEE/NASA*, pp. 153–158. Greenbelt, Maryland, IEEE, 2006.
104. Kuhn, D. R. and M. J. Reilly. An investigation of the applicability of design of experiments to software testing. In *Software Engineering Workshop, 2002. Proceedings 27th Annual NASA Goddard/IEEE*, pp. 91–95. Greenbelt, Maryland, IEEE, 2002.
105. Kuhn, D. R., D. R. Wallace, and A. M. Gallo, Jr. Software fault interactions and implications for software testing. *IEEE Transactions on Software Engineering*, 30(6), 2004: 418–421.
106. Kuhn, D. R., R. Kacker, and Y. Lei. Random versus. Combinatorial methods for discrete event simulation of a grid computer network. In *Proceedings, Mod Sim World, 2009*, pp. 83–88. Virginia Beach, Virginia, NASA CP-2010-216205, National Aeronautics and Space Administration, October 14–17, 2009.
107. Kuhn, D. R., R. Kacker, and Y. Lei. Combinatorial and random testing effectiveness for a grid computer simulator. NIST Technical Report, October 24, 2008.
108. Kuhn, D. R., R. Kacker, and Y. Lei. *Practical Combinatorial Testing*, NIST SP 800-142, October, 2010.
109. Kuhn, D. R., J. M. Higdon, J. F. Lawrence, R. N. Kacker, and Y. Lei. Combinatorial methods for event sequence testing. In *2012 IEEE Fifth International Conference on Software Testing, Verification and Validation (ICST)*, pp. 601–609. Montreal, Canada, IEEE, 2012.

110. Kuhn, D. R., J. M. Higdon, J. F. Lawrence, R. N. Kacker, and Y. Lei. Efficient methods for interoperability testing using event sequences. *Crosstalk, Journal of Defense Software Engineering*, 25, 2012: 15–18.
111. Kuhn, D. R. and J. M. Higdon. Testing event sequences. http://csrc.nist.gov/groups/SNS/acts/sequence_cov_arrays.html, October, 2009.
112. Kuhn, D. R. *Combinatorial Measurement Tool User Guide*. Available online at http://csrc.nist.gov/groups/SNS/acts/documents/ComCoverage110130.pdf, Published on January 30, 2011 and last accessed on May 14, 2012.
113. Kudrjavets, G., N. Nagappan, and T. Ball. Assessing the relationship between software assertions and faults: An empirical investigation. In *Proceedings of the 17th International Symposium on Software Reliability Engineering*, pp. 204–212. Raleigh, North Carolina, IEEE Computer Society, 2006.
114. Lawrence, J., R. N. Kacker, Y. Lei, D. R. Kuhn, and M. Forbes. A survey of binary covering arrays. *The Electronic Journal of Combinatorics*, 18(1), 2011: P84.
115. Leavens, G., A. Baker, and C. Ruby. JML: A notation for detailed design. *Kluwer International Series in Engineering and Computer Science*, pp. 175–188, Springer, New York, 1999.
116. Lei, Y., R. Kacker, D. R. Kuhn, V. Okun, and J. Lawrence. IPOG/IPOG-D: Efficient test generation for multi-way combinatorial testing. *Software Testing, Verification and Reliability*, 18(3), 2008: 125–148.
117. Lei, Y., R. Kacker, D. R. Kuhn, V. Okun, and J. Lawrence. IPOG: A general strategy for *t*-way software testing. In *14th Annual IEEE International Conference and Workshops on the Engineering of Computer-Based Systems, 2007. ECBS'07*, pp. 549–556. Washington, D.C., IEEE, 2007.
118. Lei, Y., R. H. Carver, R. Kacker, and D. Kung. A combinatorial testing strategy for concurrent programs. *Software Testing, Verification and Reliability*, 17(4), 2007: 207–225.
119. Lei, Y. and K.-C. Tai. In-parameter-order: A test generation strategy for pairwise testing. In *Proceedings of the Third IEEE International Symposium on High-Assurance Systems Engineering 1998*, pp. 254–261. Washington, D.C., IEEE, 1998.
120. Leyden, J. Symantec update killed biz PCs in three-way software prang. *The Register*, July 16, 2012, http://www.theregister.co.uk/2012/07/16/symantec_update_snafu/.
121. Luckham, D. and F. W. Henke. An overview of ANNA—A specification language for ADA. *IEEE Software*, 2(2), 1985: 9–23.
122. Lyu, M. R. *Handbook of Software Reliability Engineering*, Vol. 3. IEEE Computer Society Press, California, 1996.
123. Maker, P. J. *GNU Nana—User's Guide* (version 2.4). Technical report, School of Information Technology—Northern Territory University, July 1998.
124. Malloy, B. A. and J. M. Voas. Programming with assertions: A prospectus. *IT Professional*, 6(5), 2004: 53–59.
125. Mandl, R. Orthogonal Latin squares: An application of experiment design to compiler testing. *Communications of the ACM*, 28(10), 1985: 1054–1058.
126. Marick, B. *The Craft of Software Testing*. Simon & Schuster, New York, 1995.

127. Marick, B. Test requirement catalog: Generic clues, developer version. www.exampler.com. 1995. http://www.exampler.com/testing-com/writings/catalog.pdf (accessed September 2012).
128. Mathur, A. P. *Foundations of Software Testing*. Addison-Wesley Professional, Boston, Massachusetts, 2008.
129. Maughan, C. *Test Case Generation Using Combinatorial Based Coverage for Rich Web Applications*. Utah State University, Logan, Utah, 2012.
130. Maximoff, J. R., M. D. Trela, D. R. Kuhn, and R. Kacker. A method for analyzing system state-space coverage within a *t*-wise testing framework. In *Systems Conference, 2010 4th Annual IEEE*, pp. 598–603. San Diego, California, IEEE, 2010.
131. McCluskey, E. J. and S. Bozorgui-Nesbat. Design for autonomous test. *IEEE Transactions on Computers*, 100(11), 1981: 866–875.
132. Memon, A. M. and Q. Xie. Studying the fault-detection effectiveness of GUI test cases for rapidly evolving software. *IEEE Transactions on Software Engineering*, 31(10), 2005: 884–896.
133. Montanez, C., D. R. Kuhn, M. Brady, R. M. Rivello, J. Reyes, and M.K. Powers. Evaluation of fault detection effectiveness for combinatorial and exhaustive selection of discretized test inputs. *Software Quality Professional*, 14(3), 2012: 12–18.
134. Montgomery, D. C. *Design and Analysis of Experiments*. Wiley, Hoboken, New Jersey, 2008.
135. Musa, J. G., G. Fuoco, N. Irving, D. Kropfl, and B. Jublin, The operational profile, In *Handbook of Software Reliability Engineering*, M.R. Lyu, ed. McGraw-Hill, Chapter 5, 1996.
136. Musa, J. D., A. Iannino, and K. Okumoto. *Software Reliability: Measurement, Prediction, Application*. McGraw-Hill, New York, 1987.
137. Meyer, B. *Object-Oriented Software Construction*, 2nd edition. Prentice Hall, Upper Saddle River, New Jersey, 1997, ISBN 0-13-629155-4.
138. Meyer, B., A. Fiva, I. Ciupa, A. Leitner, Y. Wei, and E. Stapf. Programs that test themselves. *Computer*, 42(9), 2009: 46–55.
139. Myers, G. J., C. Sandler, and T. Badgett. *The Art of Software Testing*. Wiley, Hoboken, New Jersey, 2011.
140. National Aeronautics and Space Administration, Meteorological measurement system. http://geo.arc.nasa.gov/sgg/mms/Integration/dc8/mission_manager.htm. Accessed on April 23, 2013.
141. National Institute of Standards and Technology. Dictionary of algorithms and data structures. http://xlinux.nist.gov/dads/HTML/greedyalgo.html. Accessed on April 23, 2013.
142. National Institute of Standards and Technology. Test accelerator. http://www.itl.nist.gov/div897/docs/testacc.html. Accessed on April 23, 2013.
143. National Institute of Standards and Technology. Number of interactions involved in software failures—Empirical data. http://csrc.nist.gov/groups/SNS/acts/ftfi.html. Accessed on April 23, 2013.
144. Nie, C. and H. Leung. The minimal failure-causing schema of combinatorial testing. *ACM Transactions on Software Engineering and Methodology (TOSEM)*, 20(4), 2011: 15.

145. Northrop, L. M. SEI's software product line tenets. *IEEE Software*, 19(4), 2002: 32–40.
146. Nurmela, K. J. Upper bounds for covering arrays by tabu search. *Discrete Applied Mathematics*, 138(1), 2004: 143–152.
147. Nurmela, K. J. (webpage) http://www.tcs.hut.fi/~kjnu/covarr.html. Accessed on April 23, 2013.
148. Ocariza, F. S., K. Pattabiraman, and B. Zorn. JavaScript errors in the wild: An empirical study. In *Software Reliability Engineering (ISSRE), 2011 IEEE 22nd International Symposium on*, pp. 100–109. Hiroshima, Japan, IEEE, 2011.
149. Okun, V. and P. E. Black. Issues in software testing with model checkers. In *Proceedings of the International Conference on Dependable Systems and Networks (DSN-2003)*. San Francisco, California, 2003.
150. Okun, V. Specification mutation for test generation and analysis. PhD dissertation, University of Maryland Baltimore, 2004.
151. Object Management Group. *OMG Unified Modeling Language (OMG UML), Superstructure Version 2.3*.
152. Oster, S., F. Markert, and P. Ritter. Automated incremental pairwise testing of software product lines. In *Proceedings of the 14th International Conference on Software Product Lines: Going Beyond*, pp. 13–17. Jeju Island, South Korea, September 13–17, 2010.
153. Ostrand, T. J. and M. J. Balcer. The category-partition method for specifying and generating functional tests. *Communications of the ACM*, 31(6), 1988: 676–686.
154. Pezze, M. and M. Young. *Software Testing and Analysis—Process, Principles and Techniques*. John Wiley & Sons, Hoboken, New Jersey, 2008.
155. Phadke, M. S. *Quality Engineering Using Robust Design*. Prentice Hall PTR, Upper Saddle River, New Jersey, 1995.
156. Alexander, P., T. Mouelhi, and Y. Le Traon. Model based tests for access control policies. *2008 International Conference on Software Testing, Verification, and Validation*, pp. 338–347. Lillehammer, Norway, 2008.
157. Damaraju, R. *Constructions and Combinatorial Problems in Design of Experiments*. Dover, New York, 1971.
158. Rao, C. R. Factorial experiments derivable from combinatorial arrangements of arrays. *Journal of Royal Statistical Society (Supplement)*, 9, 1947: 128–139.
159. Renyi, A. *Foundations of Probability*. Wiley, New York, 1971.
160. Roux, G. k-propriétés dans les tableaux de n colonnes: Cas particulier de la k-surjectivité et de la k-permutivité, Unpublished PhD dissertation, University of Paris, 1987.
161. Qu, X., M. B. Cohen, and K. M. Woolf. Combinatorial interaction regression testing: A study of test case generation and prioritization. In *Software Maintenance, 2007. ICSM 2007. IEEE International Conference on*, pp. 255–264. Paris, France, IEEE, 2007.
162. Ramler, R., T. Kopetzky, and W. Platz, Combinatorial test design in the TOSCA testsuite: Lessons learned and practical implications, *Workshop on Combinatorial Testing (CT) International Conference on Software Testing (ICST 2012)*, Montreal, Canada, IEEE Computer Society, April 17–21, 2012.

163. Reisner, E., C. Song, K.-K. Ma, J. S. Foster, and A. Porter. Using symbolic evaluation to understand behavior in configurable software systems. In *Proceedings of the 32nd ACM/IEEE International Conference on Software Engineering-Volume 1*, pp. 445–454. Cape Town, South Africa, ACM, 2010.
164. Reuys, A., S. Reis, E. Kamsties, and K. Pohl. Derivation of domain test scenarios from activity diagrams. In *Proceedings of the International Workshop on Product Line Engineering—The Early Steps—Planning, Modeling and Managing (PLEES'03)*, Fraunhofer IESE, Erfurt, Germany, September, 2003.
165. Richardson, D. J. and L. A. Clarke. A partition analysis method to increase program reliability. In *Proceedings of the 5th International Conference on Software Engineering*, pp. 244–253. San Diego, California, IEEE Press, 1981.
166. Rosenblum, D. S. A practical approach to programming with assertions. *IEEE Transactions on Software Engineering*, 21(1), 1995: 19–31.
167. Rothermel, G., R. H. Untch, C. Chu, and M. J. Harrold. Prioritizing test cases for regression testing. *IEEE Transactions on Software Engineering*, 27(10), 2001: 929–948.
168. Salecker, E. and S. Glesner. Combinatorial interaction testing for test selection in grammar-based testing. In *Software Testing, Verification and Validation (ICST), IEEE Fifth International Conference on*, pp. 610–619. Montreal, Canada, IEEE, 2012.
169. Samek, M. A crash course in UML state machines. *EE Times*, March 2009. http://eetimes.com/design/embedded/4008247/A-crash-course-in-UML-state-machines-Part-1, http://eetimes.com/design/embedded/4008251/A-crash-course-in-UML-state-machines-Part-2, http://eetimes.com/design/embedded/4026976/A-crash-course-in-UML-state-machines-Part-3.
170. Sampath, S., S. Sprenkle, E. Gibson, L. Pollock, and A. S. Greenwald. Applying concept analysis to user-session-based testing of web applications. *IEEE Transactions on Software Engineering*, 33(10), 2007: 643–658.
171. Sampath, S., R. C. Bryce, G. Viswanath, V. Kandimalla, and A. G. Koru. Prioritizing user-session-based test cases for web applications testing. In *Software Testing, Verification, and Validation, 1st International Conference on*, pp. 141–150. Lillehammer, Norway, IEEE, 2008.
172. Sampath, S., R. C. Bryce, S. Jain, and S. Manchester. A tool for combination-based prioritization and reduction of user-session-based test suites. In *Software Maintenance (ICSM), 27th IEEE International Conference on*, pp. 574–577. Williamsburg, Virginia, IEEE, 2011.
173. Sampath, S. and R. C. Bryce. Improving the effectiveness of test suite reduction for user-session-based testing of web applications. *Information and Software Technology*, 54(7), 2012: 724–738.
174. Shi, L., C. Nie, and B. Xu. A software debugging method based on pairwise testing. *Computational Science–ICCS 2005*, pp. 55–81, Springer, New York, 2005.
175. Sato, S. and H. Shimokawa. Methods for setting software test parameters using the design of experiments method (in Japanese). In *Proceedings of the 4th Symposium on Quality Control in Software, Japanese Union of Scientists and Engineers (JUSE)*, pp. 1–8. 1984.

176. Schroeder, P. J., P. Bolaki, and V. Gopu. Comparing the fault detection effectiveness of n-way and random test suites. In *Empirical Software Engineering, 2004. ISESE'04. Proceedings. 2004 International Symposium on*, pp. 49–59. Redondo Beach, California, IEEE, 2004.
177. Segall, I., R. Tzoref-Brill, and A. Zlotnick. Simplified modeling of combinatorial test spaces. In *Software Testing, Verification and Validation (ICST), 2012 IEEE Fifth International Conference on*, pp. 573–579. Montreal, Canada, IEEE, 2012.
178. Shakya, K., T. Xie, N. Li, Y. Lei, R. Kacker, and R. Kuhn. Isolating failure-inducing combinations in combinatorial testing using test augmentation and classification. In *Software Testing, Verification and Validation (ICST), 2012 IEEE Fifth International Conference on*, pp. 620–623. Montreal, Canada, IEEE, 2012.
179. Sherwood, G. Efficient testing of factor combinations. In *Proceedings of the Third International Conference on Software Testing, Analysis, and Review*, Washington, DC, 1994.
180. Sherwood, G. B., S. S. Martirosyan, and C. J. Colbourn. Covering arrays of higher strength from permutation vectors. *Journal of Combinatorial Designs*, 14(3), 2006: 202–213.
181. Sherwood, G. B. *Getting the Most from Pairwise Testing: A Guide for Practicing Software Engineers*. CreateSpace, Scotts Valley, California, 2011.
182. Sloane, N. J. A. Covering arrays and intersecting codes. *Journal of Combinatorial Designs*, 1(1), 1993: 51–63.
183. Hiroki, S. Method of generating software test cases using the experimental design (in Japanese), Report on Software Engineering SIG, Information Processing Society of Japan (IPSJ) No. 1984-SE-040, 1985.
184. Sloane, N. J. A. (webpage) http://www2.research.att.com/~njas/oadir/. Accessed on April 23, 2013.
185. Snedecor, G. W. and W. G. Cochran. *Statistical Methods*. Iowa State University Press, Ames, Iowa, 1967.
186. Software Engineering Institute. Catalog of software product lines. http://www.sei.cmu.edu/productlines/casestudies/catalog/?location=tertiary-nav&source=10755. Accessed on April 23, 2013.
187. Software Engineering Institute. Software product lines. http://www.sei.cmu.edu/productlines/. July 19, 2012.
188. Song, C., A. Porter, and J. S. Foster. iTree: Efficiently discovering high-coverage configurations using interaction trees. In *Proceedings of the 2012 International Conference on Software Engineering*, pp. 903–913. Zurich, Switzerland, IEEE Press, 2012.
189. Sutton, M., A. Greene, and P. Amini. *Fuzzing: Brute Force Vulnerabilty Discovery*. Addison-Wesley Professional, Boston, Massachusetts, 2007.
190. Taguchi, G. *Introduction to Quality Engineering*. UNIPUB, Kraus International, White Plains, New York, 1986.
191. Taguchi, G. *System of Experimental Design, Vol. 1 and Vol. 2*. UNIPUB, Kraus International, White Plains, New York (English translations of the 3rd edition of Jikken Keikakuho (Japanese) published in 1977 and 1978 by Maruzen), 1987.

192. Taguchi, G. *Taguchi on Robust Technology Development.* ASME Press, New York, 1993.
193. Tai, K. C. and Y. Lei. A test generation strategy for pairwise testing. *IEEE Transactions on Software Engineering*, 28, 2002: 109–111.
194. Tatsumi, K. Test case design support system. In *Proceedings of the International Conference on Quality Control (ICQC'87)*, pp. 615–620. Tokyo, Japan, 1987.
195. Tatsumi, K., S. Watanabe, Y. Takeuchi, and H. Shimokawa. Conceptual support for test case design. In *Proceedings of the 11th International Computer Software and Applications Conference (COMPSAC'87)*, pp. 285–290. Tokyo, Japan, 1987.
196. Tatsumi, K. Test case design support system. In *Proceedings of the International Conference on Quality Control (ICQC)*, pp. 615–620. Tokyo, Japan, 1987.
197. Testcover.com. www.testcover.com.
198. Torres-Jimenez, J. et al. New bounds for binary covering arrays using simulated annealing. *Information Sciences*, 185(1), 2012: 137–152.
199. Torres-Jimenez, J. (webpage) http://www.tamps.cinvestav.mx/~jtj/CA.php
200. Stanford University. Computer Systems Laboratory, J. G. Udell and E. J. McCluskey. *Efficient Circuit Segmentation for Pseudo-Exhaustive Test*, 1987.
201. University of Nebraska Lincoln. Software Artifact Infrastructure Repository. Siemens traffic collision avoidance system code. http://sir.unl.edu/portal/bios/tcas.php
202. Vilkomir, S., O. Starov, and R. Bhambroo. Evaluation of t-wise approach for testing logical expressions in software. In *Software Testing, Verification and Validation (ICST), 2013 IEEE Sixth International Conference on*, Luxembourg, IEEE, 2013.
203. Voas, J. M. and K. W. Miller. Putting assertions in their place. In *Software Reliability Engineering, 1994.* In *Proceedings, 5th International Symposium on*, pp. 152–157. Monterey, California, IEEE, 1994.
204. Voas, J. M., M. Schatz, and M. Schmid. A testability-based assertion placement tool for object-oriented software. *National Institute for Standards and Technology NIST GCR 98-735*, 1998.
205. Walker, R. A. and C. J. Colbourn. Tabu search for covering arrays using permutation vectors. *Journal of Statistical Planning and Inference*, 139(1), 2009: 69–80.
206. Wang, W., Y. Lei, D. Liu, D. Kung, C. Csallner, D. Zhang, R. Kacker, and R. Kuhn. A combinatorial approach to detecting buffer overflow vulnerabilities. In *Dependable Systems & Networks (DSN), 2011 IEEE/IFIP 41st International Conference on*, pp. 269–278. Hong Kong, China, IEEE, 2011.
207. Wang, W., S. Sampath, Y. Lei, and R. Kacker. An interaction-based test sequence generation approach for testing web applications. In *High Assurance Systems Engineering Symposium, 2008. HASE 2008. 11th IEEE*, pp. 209–218. Nanjing, China, IEEE, 2008.
208. Wang, W., Y. Lei, S. Sampath, R. Kacker, R. Kuhn, and J. Lawrence. A combinatorial approach to building navigation graphs for dynamic web

applications. In *Software Maintenance, 2009. ICSM 2009. IEEE International Conference on*, pp. 211–220. Edmonton, Canada, IEEE, 2009.

209. Wang, W., Y. Lei, D. Liu, D. Kung, C. Csallner, D. Zhang, R. Kacker, and R. Kuhn. A combinatorial approach to detecting buffer overflow vulnerabilities. In *Dependable Systems & Networks (DSN), 2011 IEEE/IFIP 41st International Conference on*, pp. 269–278. Hong Kong, China, IEEE, 2011.
210. Yuan, X., M. Cohen, and A. M. Memon. Covering array sampling of input event sequences for automated GUI testing. In *Proceedings of the Twenty-Second IEEE/ACM International Conference on Automated Software Engineering*, pp. 405–408. Atlanta, Georgia, ACM, 2007.
211. Yuan, X. and A. M. Memon. Using GUI run-time state as feedback to generate test cases. In *Software Engineering, 2007. ICSE 2007. 29th International Conference on*, pp. 396–405. Minneapolis, Minnesota, IEEE, 2007.
212. Wallace, D. R. and D. R. Kuhn. Failure modes in medical device software: An analysis of 15 years of recall data. *International Journal of Reliability, Quality and Safety Engineering*, 8(04), 2001: 351–371.
213. Wang, Z., B. Xu, L. Chen, and X. Lei. Adaptive interaction fault location based on combinatorial testing. In *Quality Software (QSIC), 2010 10th International Conference on*, pp. 495–502. IEEE, 2010.
214. Weyuker, E. J. Using failure cost information for testing and reliability assessment. *ACM Transactions on Software Engineering and Methodology (TOSEM)*, 5(2), 1996: 87–98.
215. Weyuker, E. J. Testing component-based software: A cautionary tale. *Software, IEEE*, 15(5), 1998: 54–59.
216. Weyuker, E., T. Goradia, and A. Singh. Automatically generating test data from a Boolean specification. *IEEE Transactions on Software Engineering*, 20(5), 1994: 353–363.
217. White, L.J. and E. I. Cohen. A domain strategy for computer program testing. *IEEE Transactions on Software Engineering*, 3, 1980: 247–257.
218. Wikipedia. Java modeling language, http://en.wikipedia.org/wiki/Java_Modeling_Language. Accessed on April 23, 2013
219. Williams, A. W. and R. L. Probert. A practical strategy for testing pair-wise coverage of network interfaces. In *Software Reliability Engineering, 1996. Proceedings, Seventh International Symposium on*, pp. 246–254. White Plains, New York, IEEE, 1996.
220. World Wide Web Consoritum, DOM level 3 events specification, September 8, 2009. http://www.w3.org/TR/DOM-Level-3-Events/
221. World Wide Web Consoritum, Document object model conformance test suites. http://www.w3.org/DOM/Test/. Accessed on April 23, 2013.
222. Zeller, A. and R. Hildebrandt. Simplifying and isolating failure-inducing input. *IEEE Transactions on Software Engineering*, 28(2), 2002: 183–200.
223. Zhang, Z. and J. Zhang. Characterizing failure-causing parameter interactions by adaptive testing. In *Proceedings of the 2011 International Symposium on Software Testing and Analysis*, pp. 331–341. Toronto, Canada, ACM, 2011.

224. Zhang, Z., X. Liu, and J. Zhang. Combinatorial testing on ID3v2 tags of MP3 files. In *Software Testing, Verification and Validation (ICST), 2012 IEEE Fifth International Conference on*, pp. 587–590. Montreal, Canada, IEEE, 2012.
225. Cohen, M. B., C. J. Colbourn, P. B. Gibbons, and W. B. Mugridge. Constructing test suites for interaction testing. In *Proceedings of 25th IEEE International Conference on Software Engineering*, pp. 38–48. Portland, Oregon, 2003.
226. Tung, Y. W. and W. S. Aldiwan. Automating test case generation for the new generation mission software system. In *Proceedings of IEEE Aerospace Conference*, pp. 431–437. Big Sky, Montana, 2000.
227. Williams, A. W. Determination of test configurations for pair-wise interaction coverage. In *Proceedings of 13th International Conference on the Testing of Communicating Systems*, pp. 59–74. Ottawa, Canada, 2000.
228. Chateauneuf, M. A., C. J. Colbourn, and D. L. Kreher. Covering arrays of strength 3. *Designs, Codes, and Cryptography*, 16, 1999: 235–242.
229. Calvagna, A. and A. Gargantini. *T*-wise combinatorial interaction test suitees construction based on coverage inheritance, *Journal of Software Testing, Verification and Reliability*, 22(7), 2012: 507–526.
230. W. Grieskamp, X. Qu, X. Wei, N. Kicillof, and M. B. Cohen. Interaction coverage meets path coverage by SMT constraint solving. In *International Conference on Testing of Communicating Systems and International Workshop on Formal Approaches to Testing of Software*, 2009.
231. Calvagna, A. and A. Gargantini. Combining satisfiability solving and heuristics to constrained combinatorial interaction testing. In *3rd International Conference on Tests and Proofs*, pp. 27–42. Springer, 2009.
232. Calvagna, A. and A. Gargantini. A logic-based approach to combinatorial testing with constraints. In B. Beckert and R. Hähnle, eds. *Tests and Proofs, Second International Conference*, TAP 2008, Prato, Italy, April 9–11, 2008.

Index

A

D

E

F

G

H

I

J

L

M

U

V

W